Television's Teletext

Television's Teletext

Richard H. Veith

North-Holland
New York • Amsterdam • Oxford

Elsevier Science Publishing Co., Inc.
52 Vanderbilt Avenue, New York, New York 10017

Sole distributors outside the United States and Canada:

Elsevier Science Publishers B. V.
P. O. Box 211, 1000 AE Amsterdam, The Netherlands

Library of Congress Cataloging in Publication Data

Veith, Richard.
Television's Teletext.

Bibliography: p.
Includes index.
1. Teletext (Data transmission system) I. Title.
TK5105.V44 1983 384.55 82–22414
ISBN 0–444–00772–5

Manufactured in the United States of America

Contents

Preface

Four words relatively new to most of us form the setting for this book: videotex, videotext, teletex, and teletext. Actually, these look like just two words varied by the presence or absence of a final "t." During the past decade, these words have been introduced, used separately, used interchangeably, redefined, and used again to the extent that it is hard to be precise about what they mean—individually or collectively. Nevertheless, an understanding of what they generally mean will lead to a better appreciation of the unique realm of one of them, namely, teletext. So let's take the words in turn.

Videotex is a term used to describe computerized information systems aimed at the mass market. Thus the distinguishing characteristics of videotex are, usually, ease of use and attractiveness. In practical terms, videotex systems often make use of color and graphics (attractiveness) and "menu" selection as the form of interaction (ease of use). Originally, it was quite easy to describe a videotex system by saying that it connects your home television set to a computer through the telephone line. However, information systems that do not use color and do not use home television sets (or even office television sets for that matter) are also claiming to be videotex because they are easy to use and mass market-oriented. Consequently, it is still a little early to present a precise definition of videotex except as described above, and to say that another, less popular term for videotex is "viewdata."

Videotext with the final "t" has been suggested as an answer to the problem of defining videotex. The term *without* the final "t" can be used to refer only to those systems that do use telephone lines, are mass market-oriented, and use menu selection procedures for helping naïve users find information. The term *with* the final "t" can be used as the umbrella term to refer to anything that remotely resembles videotex systems, including systems that use over-the-air television as the distribution medium and are usually called teletext (defined below). Unfortunately for purists, videotex (without the final "t") seems to be gaining ground as the umbrella term. The Videotex Industry Association, for example, represents both videotex (narrowly defined) and teletext interests.

Teletex without the final "t" refers to computerized information and message systems that are built around telex machines and communicating word processors. Again, the orientation is toward mass market use. In teletex systems, pages of information are transmitted among users in normal business letter format and some of the message handling facilities are based on word processor techniques. Although teletex is primarily known in Western Europe, the major U.S. international telecommunications companies, such as RCA Globcom, Western Union International, and ITT Worldcom, are preparing to introduce teletex into the United States during the next several years.

Teletext with the final ''t,'' finally, is the subject of this book and refers to computerized information and entertainment systems that use a normal television signal to distribute the data to television sets or monitors (or perhaps even printers).

These four terms (and related terms, such as viewdata and cabletext) may be new, but the reality behind the terms is essentially basic computerized information systems, stemming from developments in computers and telecommunications during the past 20 or 30 years. However, the introduction of these terms into our vocabulary is indicative of a trend that is just beginning—the movement of such systems toward mass market use. Gradually, computer systems for information and entertainment are spreading from businesses to homes and from specialists to the population at large. This will not be a sudden change, though, because construction and use of the systems does cost money, and businesses are more willing to pay for information than the average consumer.

But of all these related systems, the one most likely to reach mass market proportions soonest is teletext, because teletext can be ''free'' (advertiser-supported) and teletext demands very little of the viewer looking at the television set. In addition, teletext systems are less expensive to create and operate than the average videotex system. As this book points out, teletext systems, in operation for several years, are now beginning to blossom. The purpose of this book is describe how, why, and what for.

One of the hazards of writing a book about a new technology is that technology is alive and sometimes seems to be growing rapidly. Statements made one year about a given system's technical aspects may be proven false the following year as the original system grows new features. Yet the world of new technology is not all that unstable or unpredictable. Ten years ago, people were talking about two-way cable television, but it is only now that many of the applications of two-way cable are beginning to be implemented. And even though the technology may be a bit different today (no one was talking about the Xerox Ethernet ® then), much of what was said about the services, and consumers' reactions to the services, remains useful. In the same way, teletext is still more talked about than seen, and as teletext grows over the next decade the technical aspects, and some of the companies involved, may vary. Nevertheless, much of what can be said now about ''how, why, and what for'' will be at least instructive, and with a little luck not far off overall, in the years ahead.

Acknowledgments

During the long process of gathering material for this book, I have been continually learning about teletext—helped by the advice and comments of a great many people. Several people whom I would like to thank specifically, but whose names I do not know, are the anonymous reviewers of the draft manuscript; their perceptive remarks helped shape the final version.

I would also like to thank the following people and organizations who provided information, photographs, or other material: Thomas N. Hastings, Chairman of ANSI Task Group X3L2.1; Graham Clayton, BBC Ceefax Service; Bob Barnes, Beston Electronics, Inc.; Michele Koensgen, Canadian Department of Communications; Dr. John deMercado, Canadian Telecommunication Regulatory Service; Frederick Simenel, Intelmatique; Eiichi Sawabe, Japan Broadcasting Corporation; Dr. Ronald Goldman, KCET (TV); Pamela J. Dorge, Keycom Electronic Publishing; Robert J. Geline, National Broadcasting Company; Netherlands Broadcasting Corporation; Red Burns and Pat Quarles, NYU/WETA (TV) Teletext Project; Gary Stein, Playcable; David Sillman, Public Broadcasting Service; Bill Sullivan, Satellite Syndicated Systems; Bill Penner, Sears Roebuck and Company; Robin Cobbey, Source Telecomputing Corporation; Kim Peters, Time, Inc.; Sheryl Rogers, Tocom, Inc.; Catherine Hurley, TV Ontario.

In addition, a number of friends and colleagues have provided information or comments, which assisted me at one time or another, that I would especially like to mention, including Gwyn Morgan and Ken Shilson of Logica, Ltd., Allan Pearson of Logica, Ltd., William Asip (who was an unending source of information), formerly of Logica, Ltd., and Darby Miller of the Hearst Corporation.

Of course none of the people who passed information along to me is responsible for any errors of fact or interpretation that I may have mistakenly allowed in the text.

Finally, thanks go to Bonnie L. Storm for providing the illustrations.

1

The Nature of Teletext

It is midnight in Chicago in September 1981, and the television set is tuned to Channel 32. The normal broadcast ends, and suddenly the screen begins displaying pages of text and graphics in basic colors, covering topics like sports results, news headlines, the weather—even a request to call the station and let them know if the bottom line on the screen is readable. It is Channel 32's "Nite Owl" service, a simulation of their experimental teletext service. Thus in late 1981 Chicago was virtually the only place in the United States where the entire viewing population of a region could get an idea of what teletext was all about (even if it was not true teletext). And yet the concept of teletext had already permeated the television industry to such an extent that the television critic of *The New York Times* could easily conclude in 1981 that by 1990 teletext will be an integral part of the television experience, bringing with it new reasons for television viewing [1].

Since the introduction of the teletext technique, which is really nothing more than "hiding" extra digital signals inside the normal television signal, a number of organizations in several countries have begun to develop services using this technique. And because these developments are relatively new, especially in the United States, the term "teletext" has been stretched considerably to describe the divergent services and applications.

For a start, we can say that teletext is a technique for bringing computerized information and entertainment to television sets without interfering with normal television broadcasts. The distinguishing feature of teletext is that digital information is inserted, or multiplexed, into unused portions of the analog television signal. Special circuitry in a television receiver is then used to read the digital information and create, on the television screen, the appropriate letters and pictures. A viewer uses a control pad, typically a remote control unit, to switch back and forth between the teletext service and the normal programming on a given channel, or even to superimpose the teletext words over the normal channel's program.

Essentially, this means that every broadcast television station in the country can become two "stations"—broadcasting the usual video signal and broadcasting a separate service using digitally coded words and pictures. Similarly, every cable television system, closed circuit television system, or restricted television system can create two "channels" in the place of one.

Of course, we are not there yet. We are just beginning. But that is what this book is all about. It is possible that 50 years from now (or perhaps 80 or 100 years from now), all television signals will be transmitted in digital form, and that this stream of digital information will actually be composed of a variety of simultaneous text, voice, and video communications. Our current techniques for slipping a few bits of data into an unused portion of an analog television signal will then be as different as Alexander Graham Bell's talking machine is from the telephones of today.

However, just as the rudimentary telephone was the start of an all-encompassing change in the way we conduct our lives at home and at work, teletext may be the as yet obscure sign of a change in our habitual methods for acquiring information and the forms in which information is received. Perhaps there will even be a net increase in the quantity of information we are able to consume. Because the television viewer can, using a button on the remote control unit, quickly switch back and forth between the program and the teletext service, the viewer can check news items, for instance, without missing a beat. Thus the viewer can acquire more information than would normally be the case while watching television.

Naturally, very dramatic changes wrought by teletext, or by the eventual fully developed versions of teletext, will be visible only after decades have passed. One of the lessons of the history of technologies is that grand premises at the birth of a technology are almost never true in the short term, while in the long term the result bears only a mild resemblance to the original idea and owes as much to all the attendant social, political, environmental, and technical developments as to anything else.

That is not to say that forecasting is foolish. Forecasting with foresight is one of the better methods of ensuring that we, as builders and creators, do more good for ourselves than harm. This book, then, will try to present the teletext picture in order to illuminate the puzzle. In that way perhaps we can see where we are going.

Teletext as we know it is primarily a British development, even though the technique was being tested at about the same time in a number of countries, including the United States and Japan. As Chapter 2 explains, teletext got its start in England in the early 1970s and seems to be on the verge of flourishing. Subsequently, teletext systems have been developed and teletext services inaugurated in other Western European countries, North America, and Japan. In the United States, teletext has taken several different forms. There are close to a dozen different systems being tested in a variety of environments. Teletext on a cable television system can be quite different from teletext broadcast by a television station. Digital information has even been multiplexed into FM radio signals, thus creating yet another version of teletext.

It might be best, then, to describe the teletext technique in more detail before sketching the variations and the applications that are the subjects of later chapters.

VBI Teletext

There is more than one way to "hide" digital information inside a normal television signal, but the most common method employed in teletext systems uses that portion of the television signal called the vertical blanking interval (VBI). A television picture is created on a television screen as a series of closely spaced lines; an electron gun within the picture tube shoots a stream of electrons systematically across the inside face of the picture tube. But the electron gun needs time to jump from the bottom of the screen to the top in order to begin the next field.[1] During this time, scan lines are being broadcast but they are more or less unused. This forms the vertical blanking interval and is physically located at the top of the television screen above the viewable limit of the screen (see Figure 1.1). In fact, the number of viewable lines may vary from 483 to 499 (depending upon the television receiver) out of a total of 525 lines in the U.S. television system.

The vertical blanking interval, then, is composed of 21 lines of a television picture field that can be put to use. In the United States, for example, lines 1 through 9 form the vertical synchronization pulse, and lines 17 through 19 can carry test and reference

Figure 1.1. The unused scan lines of the vertical blanking interval.

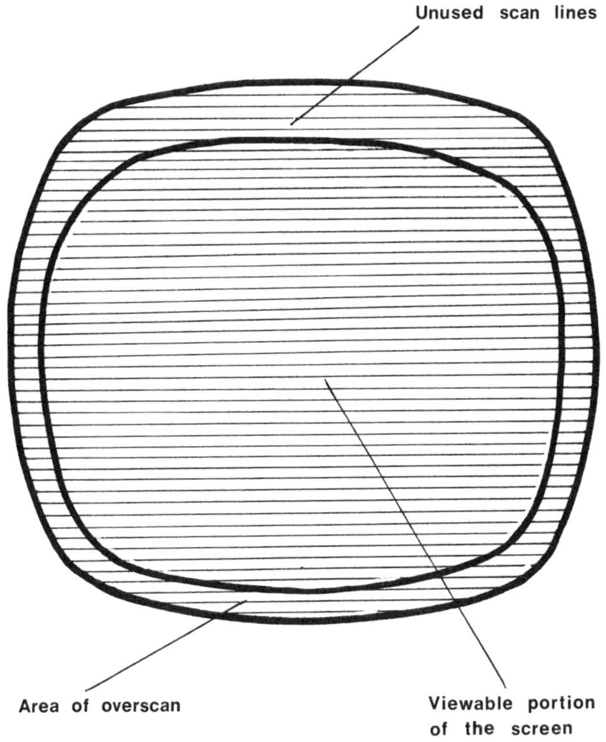

[1]Television normally employs interlaced scanning. If all lines are numbered sequentially, the odd-numbered lines are created first (the odd field); then the beam returns to the top and creates the even-numbered lines (the even field). There are therefore two fields for every complete frame, and there are 30 frames per second in U.S. television.

signals (see Table 1.1). Line 21 is currently used to carry the data to create captions for the hearing-impaired, which is discussed at greater length in Chapter 3. Because lines 10 through 13 cannot usually be used for teletext without causing interference with the picture on older television sets, lines 14 through 21 are really the focus of all the teletext turmoil.

It is worth mentioning, because the subject will come up again in later chapters, that television formats are not uniform around the world. In Western Europe, for example, there are 625 scan lines in a television picture, with 563 to 587 viewable lines. In Great Britain, teletext can technically occupy lines 7 through 22 and lines 320 through 335 (the second field) even though only a relatively few lines are currently being used. (In the United States, we tend to number each field separately, so that we can speak of using line 16, for instance, in both fields, rather than line 16 and line 278.)

Within these few lines it is possible to multiplex digital information, with the television signal alternating between the black level and near the peak white level to indicate the presence of 1s and 0s that comprise the digital information. The several proposals before the Federal Communications Commission in 1981 suggested how bits of data (i.e., 0s and 1s) could be placed on these vertical interval lines. One proposal, for example, called for 296 bits per line, or 37 bytes of eight bits each. Another specification suggested 288 bits, or 36 bytes of eight bits each.

Although the number of VBI lines being used, and the quantity of data per line, are both rather small, a large amount of textual and graphic information can actually be transmitted within a small amount of time. Typically, it takes less than one-fourth of a second to broadcast the digital information to create a full page or screen of text.

Table 1.1. Allocation of the First 21 Lines of the Television Picture in the United States (both fields)

Lines	Allocation
1	Vertical synchronization pulse
2	"
3	"
4	"
5	"
6	"
7	"
8	"
9	"
10	Unused, but not technically suitable
11	for current teletext
12	"
13	"
14	Teletext (experimental)[a]
15	"
16	"
17	Multiburst test signal (field 1)
	Color bar test signal (field 2)
18	Composite test signal
19	Vertical interval reference signal
20	Identification signal (field 1)
21	Captioning for the hearing impaired

[a]The Federal Communications Commission has proposed that lines 14–18, 20, and 21 be available for teletext purposes.

One of the chief characteristics of broadcast teletext, on the other hand, is that the systems do not usually involve more than a few hundred or so pages of text at any given time. This is a direct result of the fact that we are dealing with a broadcast system—there is normally no return channel from the viewer to the television station. Instead, the television station is broadcasting *all* of the teletext pages in a continuing cycle. If the viewer is looking at page 9, for example, and presses 30 on the control pad, page 30 will not appear on the screen until it comes around in the broadcast cycle. If page 31 had just been broadcast, the viewer would have to wait until the cycle ended, restarted, and reached 30. (This is something of an oversimplification because teletext systems do not have to broadcast the pages in strict sequence and can in fact send out the most popular pages at frequent intervals.) Obviously, if the total number of pages is kept small, the viewer would not have to wait very long even in the worst-case situation. Current teletext systems sending out about 100 pages usually take 22 to 25 seconds to complete a cycle, although the time can vary by the amount of data on each page.

Teletext is not quite as rigid and slow as this might sound. A number of techniques are being developed to increase the speed and quantity of information delivery via teletext. As mentioned above, one simple technique is to insert popular pages into the broadcast cycle out of order. The index page, for example, can be broadcast every three or four seconds, ensuring that no one would wait longer than that to get the index page, and most viewers would perceive almost instantaneous access. A second technique is to build page memory into the television set and logically link related pages together at the teletext computer. Then if a viewer chooses a sports page, for instance, the related sports pages could be picked off automatically by the home television set for storage and instantaneous access later, while the viewer is still reading the first page. In other words, the television set ''anticipates'' the viewer's next decision. A third possible technique is to use various forms of data compression to reduce the amount of digital information transmitted. In that case the decoder in the television set would have to rebuild the digital information from the compressed data.

Finally, teletext does not have to remain a broadcast, or one-way, system. A feedback channel can be provided from the viewer to the teletext computer so that the viewer can actually control the broadcasting of the pages in an interactive manner similar to most other computerized information systems. One way to do this is to call the computer by Touchtone telephone and use the telephone's number pad to send instructions. This had already been tested by KSL–TV in Salt Lake City (see Chapter 3). Another way to accomplish the same thing is to use a cable television system for the return link or for both the transmission and return link.

The concern about the number of pages or screens of text and graphics to transmit has affected the way in which teletext services have been developed. The concept of a ''magazine'' of pages has been introduced, with a magazine rarely containing more than 100 pages if the transmission of 100 pages would result in a worst-case waiting period of about 25 seconds. At any given time, the magazine will probably contain 100 pages or fewer, but it is quite possible, using existing equipment, to quickly and automatically replace pages in the magazine with new pages or with pages from a library of existing pages. The entire magazine can even be replaced automatically if such commands have been stored in the teletext computer.

Some teletext systems can introduce even more variety into the service with a distinction between normal pages and rolling pages. A normal teletext page will stay on the screen until the viewer makes another choice, switches back to the television program, or turns off the set. However, a set of rolling pages will do just that—once the viewer begins looking at the first page of the sequence, succeeding pages will appear according to the timer set by the teletext operator. In addition, some systems permit the text to scroll up or across the screen, and some permit a certain amount of animation.

Subtitles

Because the teletext technique is essentially only a means of placing digital information into the television signal, that digital information can be used for more than creating the text and graphics pages of a teletext service. The data can also form the basis for sending subtitles, either for the hearing-impaired or for different language groups. The subtitles can appear anywhere on the screen and can be displayed in various colors, with colors perhaps keyed to characters of situations. Three or four different streams of subtitles, or captions, each in a different language if desired, can be broadcast as part of the same television program using current teletext systems. The viewer is able to select the appropriate set of subtitles.

The line 21 system for captioning in use in the United States is a limited form of teletext. It is discussed further in Chapter 3.

Extensions

Building upon the basic teletext system, we can make additional features part of a teletext service. For example, it is possible to add security codes to teletext pages so that only one group of television sets, or even one set, can display the page or pages. This could be applied in a business or professional environment as another means of electronically delivering messages to a group of constituents either on a group or personal basis [2].

Another application of codes would be to use them as a subject classification system. The decoder in the home or office television set would recognize the codes and, upon command, request and store only the pages from the selected category or categories.

A feature already in place in some teletext systems is an ''alarm clock'' page. This is actually transmitted every minute, but the viewer who turns to it does not see anything (except the normal television program) until the bits representing the real time in the teletext page match the time preset by the viewer at home. In practice, the viewer turns on the television set, selects the alarm clock page and specifies a time, then continues to watch the television program on that channel. When the ''alarm goes off,'' the television set will begin to display teletext.

Beyond the current level of teletext graphics, teletext can be used, as has been demonstrated, to transmit enough information to create a full-color photographic quality picture on the television screen—in other words, digital television as still photographs or as a slowly changing sequence.

And beyond text or pictures of any kind, the digital information in the vertical blanking interval can be any type of data that can be fed into a computer, or more likely a microprocessor, in the receiving set. These data might be processed by the microprocessor or they might themselves be a computer program that will be executed by the processor in the receiver. The possible uses for this kind of digital distribution include not only mundane business applications but also new forms of games. For example, a sort of citywide bingo could be played where instead of numbers being called, data are transmitted via teletext. And instead of cards, there would be miniature computer programs in the receiver to accept and process the data and even notify the player of a win.

Another possibility is to use the broadcast digital information to drive printing machines. In fact, this was one of the considerations that led to the development of teletext in the first place. This capability provides the equivalent of a broadcast telex service, but it can also be tailored for such special services as printing coupons in the home, to be used in conjunction with advertising on the teletext service or on the video program. The idea of using a broadcast signal to carry encoded data within it destined for printing devices has also been accomplished in radio broadcasting. The digital information is inserted within an FM subcarrier. This technique has its own name: FM SCA (*s*ubsidiary *c*ommunications *a*uthorization). The current most common use of SCA though has nothing to do with distribution of data; instead, SCA is used to send Muzak and similar background music to subscribing businesses.

Finally, beyond using the vertical blanking interval itself, it is possible to devote all of the usable lines of a television signal to teletext, creating what has been termed "full-channel teletext" or "full-field teletext." This is somewhat of an aberration of teletext because there is no longer any normal video signal left. Instead, a televisionwide signal is being devoted to data transmission, with the provision that the data are still being multiplexed into a television signal "envelope." One advantage of this is that a previously unused television channel can become a means for distributing information on a pseudodemand basis. For example, in France a full-channel teletext system is being used to broadcast stock market reports to specially equipped television sets. The viewers, or users of the service, feel that they are interacting with an on-line information system because they see stock listings only as they make requests on a control pad, and the response time is very fast. But the appearance of interaction and extremely fast response time results from the capacity of the televisionwide signal. A great deal more information can be distributed per second using 200 or 300 television scan lines than by merely using one or two lines.

In the United States, the newer cable television systems with their extra channels are already exploring the use of full-channel teletext for distributing digital information to users. In this way the digital information can be easily carried by cable television equipment designed to handle television signals. And over-the-air subscription television stations and the new low-power television stations can broadcast full-channel teletext (probably for business purposes) when not broadcasting their premium fare, or pay TV.

Why Teletext?

The teletext technique has been developed, tested, and put to use. But why exactly does teletext appeal to people and why will it grow?

There are several answers to these questions, depending upon the group involved—television station, cable system, publisher, consumer—but almost all the answers will require at least a decade before they may be borne out. The present stage of teletext resembles the early days of color television. The picture was prettier, but it still took a while for the new television sets to penetrate the market and for advertisers to want to pay the additional money required for the new type of programming. But once the enhanced service approached wide visibility in the marketplace, consumers expected it and were happy with nothing less, even if they still owned black and white television sets. In 1960 the number of color television sets sold in the United States was less than

2 percent of the total; by 1970 the number of color sets was slightly under half the total sold; and by 1979 almost twice as many color sets as black and white sets were sold. (Yet today, about 30 years after the first color broadcasts, some 16 percent of all television households in this country still do not have color television sets.) Although the introduction of teletext-equipped television sets will not necessarily strictly parallel the introduction of color-equipped television sets, we may still draw some important conclusions, namely, that teletext will become widespread as the television stations (or networks) and the set manufacturers begin actively to promote the idea in every possible way, based on the belief that teletext operations will increase station revenue and set sales respectively.

From the station operator's point of view, a teletext service effectively gives the station two channels in place of one. Both channels can be supported by advertising, thus increasing the amount of station revenue. In practice, the advertising on teletext may be in a form unlike that on the normal channel, but it is still advertising. There are other uses, though, for teletext as mentioned above. The station operator could elect to engage in special services (e.g., broadcasting data to special interest groups, perhaps on a subscription basis). Or the station operator could decide simply to lease the space—the teletext lines in the vertical blanking interval—to other organizations that want to exploit the technique.

Cable television systems do not have quite the same use for teletext that broadcast television stations do. A cable system not only has dozens of channels, all of which might carry data in the vertical blanking interval, but also may already have text services appearing on normal channels as scrolling or crawling text. Regarding the latter, the teletext technique does offer two advantages. First, teletext brings selectivity. The viewer can look at an index and select the desired page without waiting for text to scroll by until the desired item appears. The second advantage is that a normal scrolling text service does occupy a full television channel, but a teletext service can be squeezed onto an existing channel, which can be quite an advantage on cable systems already fully occupied.

Cable television is also unlike broadcast television, with respect to teletext, in the availability of alternative techniques. If a cable operator intends to offer an information service or text-based entertainment service that requires the sending of digital information to homes, it is possible to use data transmission methods that do not involve inserting the data into a portion of a television signal. The data can be modulated and transmitted down a very narrow part of the frequency spectrum available on a cable in a manner almost identical to that for using telephone lines. A number of cable television systems have been doing this for business customers for quite some time. Yet teletext still offers an advantage in that, as in the cable system, it is designed for a television signal environment. Thus teletext can be added to a cable system without much if any change to the equipment throughout the system that has been designed to handle television signals. Some cable systems, in fact, have considered using the teletext technique of inserting data into the vertical blanking interval as a means of sending control signals to individual homes to authorize access to certain channels or programs (e.g., for pay TV).

From a publisher's point of view, teletext offers another distribution medium for electronically stored text and graphics. The publisher who has information suitable for a teletext service, such as a local newspaper, might decide to prepare news items, weather, sports information, or classified ads for teletext distribution and then sell that information to the local television station that wants to run a teletext service for the advertising revenue but does not want to get involved in a large newspaperlike editorial effort.

The customer's point of view, of course, is somewhat different, because traditional broadcasting generates neither income nor bills. The most likely appeal to the average viewer is the ability to immediately acquire information for free, or to engage in some entertainment service delivered via teletext. Surveys of the teletext viewers in Great Britain indicate that, while teletext is popular, there is no surprising new reason why viewers turn to the teletext pages. Most often viewers are looking for news briefs, sports scores, program listings, weather reports, and schedules for local events—the same sort of information we are accustomed to receiving as brief announcements between shows [3].

There is one more, minor, player in the teletext arena: the long-distance distributor of digital information. Because television signals are distributed nationally via satellite and cables, data have been inserted into the vertical blanking interval of these signals merely as an alternative to leasing space from a common carrier. For example, data inserted into the signal of WTBS–TV, the superstation from Atlanta distributed to thousands of cable systems via satellite, are also distributed and "captured" by cable systems with the proper equipment. Although not necessarily dramatic in impact, this particular technique is nonetheless a valid variation of the normal use for teletext.

All the reasons for teletext to grow, such as the increased advertising revenue and being an additional medium for distributing data, assume the availability of television sets able to decode and display the pages of text and pictures. But it is not clear what impetus will convince both TV set manufacturers (and adapter manufacturers) and television stations (and cable companies) that teletext should be pursued now. The wary station operator knows that the vertical blanking interval will still be there for years to come. The space can always be exploited then, when enough teletext-equipped television sets have been purchased, the owners lured by the teletext offerings of some more adventuresome television station.

The solution to this dilemma may lie in several directions. First of all, teletext may gain entry into the marketplace on the shoulders of special services. These are information services that use a station's vertical blanking interval to broadcast information to a specific group of clients. The clients, either individually or as an organization, are willing to pay for the service because of the value of the information distributed and because teletext can be a cost-effective distribution medium. Second, individual companies who own television stations may decide to plunge ahead and introduce a teletext service hoping that this early entry will result in later domination of the market, and perhaps provide a background of expertise (and a data base) that can be marketed to other television stations and cable systems. And third, the television set manufacturers could conclude that mass production of teletext-equipped sets, to be sold at a price just slightly above that of a normal set, would result in a jump in the number of television sets sold.

Projections

Given the infancy of teletext in the United States at the beginning of the 1980s, there are some remarkably rosy projections for teletext. One research firm claims that by 1985 all new television sets will have integral teletext circuitry, and that by 1991 almost all households will have a television set able to decode teletext [4].

A less optimistic projection was reported by two San Diego State University researchers who based their findings on a Delphi study involving a panel of teletext experts [5]. In the Delphi procedure of recirculating the forecasts several times among the experts, an eventual consensus was reached regarding some of the forecasts, namely, that by the year 2000 somewhat over half of all U.S. homes will be able to veiw teletext. But the experts did not think that teletext would be an especially profitable venture for station operators and information providers. Moreover, the panel was collectively unsure of such items as whether incompatible teletext systems will continue to exist, what the role of closed captioning for the hearing-impaired will be in teletext development, and whether regulatory problems will hamper teletext growth. The panel did suggest that services for special interest audiences might be a teletext staple.

The Institute for the Future has also conducted a similar survey that is slightly more conservative. The institute convened meetings of people familiar with teletext, discussed the possible futures for teletext (and videotex), then followed the meetings with a questionnaire to participants. Based on 75 responses from people who, for the most part, classified themselves as either experts or very familiar with teletext, the survey showed that only 35 to 40 percent of U.S. households will have teletext TV sets by the year 2000 [6]. Asked to forecast the most popular types of teletext services, the survey participants predicted that information retrieval would be the most popular, followed by games/entertainment and data processing.

One of the possible first steps for teletext that remains largely untested is program-related teletext. Gary Arlen, editor of a videotex/teletext newsletter, suggests that there is a great potential for such things as an electronic background file for specific sports events [7]. During a football game, for example, teletext pages could carry a wealth of game statistics and background information that the home viewer could page through at will; the data could superimpose over the video so the viewer does not have to miss a thing.

Uncertainty

The future for teletext, despite the long-term projections, is by no means clear in the short term. There are, for example, a large number of unresolved legal questions. While most of these questions might apply to the general case of computerized information systems, one question is definitely a teletext problem: who owns the vertical blanking interval? If a broadcast television station is sending out teletext, can a cable operator pick up that signal, erase the teletext, and insert the cable system's own teletext service? Or if a television station's signal is being distributed nationally and is carrying teletext, can the distributor, such as a satellite service operator, replace that teletext with another stream of digital information to be distributed?

The second question above has already gone to court. In early 1981, Chicago's WGN–TV, a superstation, began broadcasting a limited amount of teletext data in its vertical blanking interval. The station owners said that they intended to receive the data at a cable television system they own in Albuquerque, New Mexico. However, United Video is the company that distributes WGN nationwide via satellite, and United Video had other plans for that vertical blanking interval. The WGN teletext data were stripped off and replaced with other digital information that United Video wanted to distribute to cable television systems. WGN–TV sued on the basis of copyright infringement, and United Video countered that the teletext data were not part of the television programming but were instead an entirely separate service that would have to be considered as such, and paid for as such, if United Video was to distribute it. In October 1981, the U.S. District Court for the Northern District of Illinois dismissed WGN–TV's complaint, indicating that the copyright was not being infringed. In August 1982, an Appeals Court reversed that decision and said that program-related data must be carried. The general question of ownership of the vertical blanking interval was not decided, however, and remains to plague the initiators of teletext services.

Another uncertainty, also involving satellites, is the possible effect of direct broadcast satellites—satellites that can broadcast television signals directly to small home antennas. Admittedly, any large-scale effects of direct broadcast satellites are years or decades away, and a great deal can be done with teletext in the interim.

The remainder of this book, then, describes in more detail the teletext systems and services that are appearing. Chapter 2 presents and discusses the teletext systems outside the United States. Some of these systems predate the teletext systems in the United States and have profoundly affected teletext development here by being both examples of what to do and what not to do.

Chapter 3 outlines the broadcast teletext systems that have appeared in the United States—in Los Angeles, Chicago, Washington, D.C., and elsewhere. Initially, most of these systems were constructed using existing technology from foreign systems.

Chapter 4 takes a more detailed look at the cable television industry's involvement in teletext. As has been mentioned, teletext on cable can be quite a different matter, with possibilities that do not exist for broadcast television.

In Chapter 5 the difficult question of standards is addressed. The Federal Communications Commission has proposed that television stations not be restricted by law to any single standard method for encoding and transmitting teletext but instead use any system of their own choosing. Although this is an admirable attempt to allow marketplace competition to pick the best system, what has actually happened is that an industry organization (the Electronic Industries Association) has put together a voluntary standard that the major broadcasters and major television set manufacturers seem to accept. However, other standards do exist (most notably the line 21 system), the standard is only voluntary, cable television systems do not need to follow broadcast standards, and there may be good business reasons for installing a ''nonstandard'' system. Thus the technical questions and features behind the standards continue to be relevant topics.

Chapter 6 relates the subject of teletext to the systems called videotex. As the chapter explains, videotex systems are really nothing more than interactive computerized information retrieval and transaction systems. The only relatively new aspects of the computer systems called videotex are the use of color and graphics and a mass market orientation. Unlike videotex, teletext stands out, then, as indeed a new technique that has yet to be exploited.

The seventh chapter steps back from all the fine details of competing systems and conflicting experiences and looks at the long-term effects for teletext in general. Because teletext is, at least now, primarily a text service (i.e., text appears on the television screen), we can eventually expect some changes in our attitudes toward the television set. Moreover, teletext calls upon the viewer to take an active role in selecting pages, instead of passively watching the screen. And if, as is likely, teletext develops as a news and information service, there is the possibility that television will outgrow the complaint that it provides only news capsules and not news in depth. Television will be able to rival newspapers in quantity and perhaps even in content, including the classified ads, with the added distinction of being able to update—to change or replace information on the teletext channel almost instantaneously. The chapter then summarizes the development of teletext and suggests possible and probable roles for teletext in our society.

In the past hundred years, communication has become *tele*communication and then computerized and digitized telecommunication. Television, which began to be a social force only within our lifetime, is now in the beginning stages of a slow transformation that is part of the longer trend in computerized telecommunications. Computers and digital telecommunications are now taking over part of the television signal and will subsequently become part of the television effect. In the future, television may *be* teletext.

References

1. O'Connor, John J., The Home Screen's New Face, *The New York Times,* September 6, 1981, p. D19.
2. Morgan, Gwyn, Inside Teletext, *Hobby Electronics,* September 1980, pp. 12–16.
3. Reply Comments of the United Kingdom Teletext Industry, submitted to the Federal Communications Commission, July 21, 1981, p. 19.
4. Separate Studies Predict 1990 Videotex Markets of $5 Billion or $9 Billion, *Videotex Teletext News,* August 1981, p. 6.
5. Vermilyea, David and Wylie, Donald, Teletext in the Year 2000: A Delphi Forecast, *IEEE National Telecommunications Conference,* 1980, pp. 23.4.1–23.4.4.
6. Tydeman, John (project director), Teletext/Videotex Questionnaire Preliminary Results, Institute for the Future, July 21, 1981.
7. Arlen, Gary, The Revenue Potential of Teletext, *View,* June 1981, p. 91.

2

In England and Elsewhere

In the late 1960s, as talk began of ''wired cities'' and ''wired nations'' to express the notion that many new forms of computer telecommunications were imminent, investigations also began into the uses of portions of broadcast signals for transmitting digital information. In England, Japan, and the United States, to name a few countries, the start of teletext became visible. But it was the English system that reached maturation first and became the guide for other systems, and therefore it is the English experience that will be described first in this survey of teletext services in other countries. (Teletext in the United States is the subject of Chapters 3 and 4.)

England

Sometime during the latter part of the 1960s, engineers at the British Broadcasting Corporation's Designs Department began what Gwyn Morgan terms the ''first rumblings of teletext'' [1]. At about that time, broadcasters in a number of countries were working on methods to utilize the vertical blanking interval for various control signals. At the BBC, aware of the control signal applications, researchers were generally looking for a way to deliver information to home-based printing devices as a service to the hearing-impaired. The primary problem was the printer, because the BBC wanted a device that would be quiet, unobtrusive, free of maintenance, and inexpensive. The transmission technique was already developed, namely, coding the data into the vertical blanking interval in the same way that the control signals are transmitted.

In 1968, as computer terminals were being built using semiconductor technology to generate characters on the screen, it became apparent that this might solve the problem of finding an inexpensive, easily operated printer. Instead of a hard copy printer, the characters could be displayed on a television screen using the semiconductor technology that was becoming available and affordable. Having accepted the idea that the digital information would appear on a television screen equipped with character-generating circuitry, the BBC engineers returned to the examination of the transmission side of the project.

One of the alternatives to inserting the data in the vertical blanking interval was to impress them onto a subcarrier in the television signal, and this method then underwent testing in the BBC laboratories. Eventually though, the experimenters concluded that use of the vertical blanking interval was the better method, creating less disturbance to an existing television signal.

At the same time, the BBC was actually thinking in terms of two different services. The first service would be subtitles or captions for video programs, visible only on television sets with the appropriate decoder. The working title for the service was "teletitles." The second service using the teletext technique would be the delivery of a full screen of text, again to sets with the appropriate decoder, with the tentative title "teledata." Because both proposed services were to use the same teletext technique, it soon seemed advantageous to design the two systems to use the same decoder. Before long, the two systems were essentially one and became known as "Ceefax."

By 1972 the BBC was ready to announce the Ceefax project and to encourage some industrywide support for the system. Obviously, the success of the Ceefax project would depend upon the willingness of television set manufacturers to produce sets with internal teletext decoders, or to produce external set-top adapters. In the same vein, the support of the independent television stations (ITV) would also be important, because the independent stations had been concurrently working on their own version of a teletext system. A committee of representatives from the BBC, ITV, the semiconductor industry, and the television manufacturing industry was established to design a suitable system standard. During 1973 the committee discussed and tested all the variable features of the proposed techniques, including different numbers of rows and columns of text, different methods of coding the text so that it would be displayed at the proper location on the screen, different techniques for creating the digital codes themselves, and a range of data rates (i.e., the amount of digital information that can be transmitted in a given moment). The upper limit on the data rate is more or less dictated by the bandwidth of the television signal, and data rates from 3.5 megabits per second to over 6 megabits per second were tried. (In England, a television channel is 8 MHz wide, with the vision portion being 5.5 MHz.)

The result of the committee's year or so of effort was the 1974 *White Book* giving the public specifications for teletext service in the United Kingdom. This specification established the format of the teletext page (i.e., a maximum of 40 characters per row and a maximum of 24 visible rows, with an additional 8 nonvisible rows for later enhancements or expansion), the coding techniques for packaging the character data (i.e., an extra 5 bytes per row for addressing information), and the data rate of 6.9 megabits per second. The data rate is actually a burst rate, meaning that when data are being transmitted, they go at that speed, but because there are also moments when no data are being transmitted, the effective data rate is considerably less.

In addition to the technical committee's work, a government committee on the future of broadcasting also examined and endorsed the teletext concept. Thus with careful preparation and cooperation, the introduction of teletext services seemed assured.

During the years that this was taking place, the British Post Office was developing its own videotex system, later to be called "Prestel," also to utilize home television sets as the display device, although using telephone lines to connect the sets to a computer. Because, again, it seemed that similar coding standards would benefit all participants, the British Post Office agreed in 1974 to adopt the teletext specifications for format, colors, character sets, and display features. (Prestel is discussed again in Chapter 6.)

Almost immediately, the BBC began broadcasting a nascent Ceefax service, and the independent stations began their own teletext service called "Oracle" (see Figure 2.1). The name Ceefax may have come from "BBC-facts" crossed with "see facts," while the name Oracle was formed from Optional Reception of Announcements by Coded Line Electronics. Actually, the BBC had been broadcasting experimental teletext since mid-1973 using a rather cumbersome page creation system whereby a journalist typed the pages on a machine that created a punched paper tape that was then fed into a memory device.

The first two years of Ceefax and Oracle were a trial period, approved by the government in late 1974. During 1975 and early 1976 the BBC and ITV were broadcasting teletext on a regular but limited basis. At the BBC, for example, owing primarily to the physical limitations of the teletext equipment, the Ceefax service contained only about ten pages of text (although pages might change throughout the day) on BBC1 and a similar number on BBC2. The subject matter tended to revolve around news, reports of traffic conditions, and stock market reports.

This was also a period of continued technical testing of the teletext technique itself and of the announced specifications. During this time and over the next several years as well, BBC and ITV engineers measured over 25,000 parameters affecting teletext trans-

Figure 2.1 Sample Ceefax page, showing financial headlines. *(Courtesy of the BBC's Ceefax Service.)*

missions [2]. The ITV group, for example, sampled the effects on teletext reception attributable to electromagnetic noise, ghosting, cochannel interference, and the like. In general, the results indicated that over 94 percent of the sample homes could receive teletext without difficulty. Some of the measurements were also conducted on wired, or cable, television systems (e.g., in apartment complexes), because about 10 percent of U.K. homes receive their television that way; and similar positive results were found except for some HF (high-frequency) distribution systems. But a few things did need correcting. In the early days of ITV's teletext transmissions, an audible buzz was heard on television sets in certain areas until ITV established stricter controls over the data amplitude.

Because of the trial nature of the teletext services, it could still be said in mid-1976 that there was no widespread use of teletext by consumers, and most people's experience with teletext, if any, was an occasional glance at a teletext page on a television set in some public place, such as a television rental store.

The combination of continued testing and experience with a live service led to the publication of a second specification in 1976. This second set of standards permitted some features to be added that would improve the teletext display, such as allowing graphic representations to contain colors adjacent to each other. This was an improvement over the previous system of leaving a blank spot on the screen where a color code had been inserted to change the display from one color to another. The government gave the final approval to teletext on November 9, 1976.

According to Colin McIntyre, the BBC journalist responsible for the beginnings of the Ceefax service, the concepts guiding the service's content and structure were patterned after radio [3]. Radio, like current teletext, reaches only one of the senses, and the radio format of capsulized information seemed most applicable to teletext. In addition, the BBC saw teletext as an extension of its ongoing information gathering and dissemination via radio and television. Thus teletext at the BBC would not be an entirely new service seeking to distribute types of information that the BBC had not previously been concerned with. Instead, the Ceefax editors would make use of the existing, and extensive, news and information gathering activities of the BBC. The distinction between content or editorial orientation and the teletext technique remains, however, and the BBC brand of teletext in the early years, while often cited as the prime example of what teletext means, is certainly not indicative of the range of possible teletext services, as will be seen throughout this book.

The growth of teletext in the United Kingdom following the trial years may turn out to be indicative of teletext growth in other countries, even though there are arguments that other countries will move faster. On the one hand, we might say that teletext growth in England, where over half the population rents or leases television sets rather than buying them, will ultimately be faster than in countries where the population primarily purchases sets. On the other hand, initial growth in England was hampered by the lack of a ready supply of the custom-designed semiconductor circuits, or chips, used in manufacturing the decoders, while other countries may be spared that problem. The net result of both positive and negative factors may be that the experience in England will not be unlike the introduction of teletext in other industrialized countries.

At any rate, the number of teletext-equipped television sets sold in Great Britain from 1974 to 1978 totaled fewer than 5,000. But in 1979, the first year after teletext sets began being manufactured on a production line basis, the number of teletext sets in use rose to 40,000. The following year that number more than tripled, by the end of 1981 there were over a quarter of a million, and by late 1982 there were over a half million sets able to receive Ceefax and Oracle. And that number is currently growing at about 8,000 sets sold per month. This represents the beginning of rapid growth, although realistically teletext is still a marginal entry in the British television scene because teletext sets account for only about 3 percent of the total number of television households, and a 1981 survey revealed that about half the population had not yet heard of the names Ceefax and Oracle [4].

Naturally, the growth of the number of teletext sets in use sparked a number of viewer surveys to determine the characteristics of the teletext audience. At least three such surveys have been undertaken—one by Communications Studies and Planning, Ltd., a British research firm; one by Philips Video, a manufacturer of teletext equipment; and one by the BBC itself [5]. All three surveys concluded that teletext viewers belonged to upper-income categories (e.g., professionals and managers). One of the surveys explained that the typical teletext viewer is young and affluent and already accustomed to watching over 30 hours of television a week.

Two of the surveys attempted to find out just how much time is spent watching the teletext service. Both concluded that nearly two hours per week are spent reading teletext pages. In practice, the two hours are composed of several teletext viewings per day, where four or five pages might be read at a time. One survey suggested that the teletext services are looked at, on average, over ten times per day for short periods.

As might be expected, teletext viewing follows usual television viewing patterns over the course of a day, with light use during the morning, a slight rise in the afternoon, and the bulk of the viewing (or reading, in this case) in early evening.

Regarding the most popular pages on the teletext services, the overwhelming favorite was news for BBC1 and ITV's Oracle. However, BBC2, which has a different style of teletext, is watched more for the teletext games and puzzle pages. Program schedules ranked second in popularity on BBC1, while sports news ranked second on Oracle, and the newsflash facility was the second favorite on BBC2, rivaling the game and puzzle pages on the latter network. (The newsflash facility gives the viewer the option of choosing to see teletext newsflashes cut into the normal video; thus a viewer continues to watch the normal program and sees the newsflashes only as they are transmitted, often with a note that the complete story can be found on another teletext page. A viewer who does not wish to see newsflashes, of course, can elect not to.) Other favorite pages were financial information, weather and travel news, and farm and garden information on BBC1; consumer news on BBC2; and weather, travel news, and financial news on Oracle. When asked about what other items they would like to see, at least a third of the survey sample mentioned local event information and local news.

Another survey, a small project involving interviews of some 30 U.K. teletext users, was conducted in the summer of 1981 by William Herring, a graduate student at New York University [6]. Although the sample size—the number of people interviewed—was quite small, the sample showed some similarity with national averages (e.g., 65 percent of the sample rent their television sets, 75 percent were male, and the most popular teletext pages were news, weather, sports, and entertainment information). Interestingly, the number of times per day for which the interviewees said they switched to teletext hovered about two to four, even though several said they switched to teletext only once per day, and several others claimed to watch teletext eight to twelve times per day. The average length of any one teletext "session," according to most of the interviewees, was anywhere from 1 to 15 minutes (seven people said 1 to 5 minutes, eight people said 6 to 10 minutes, and nine people said 11 to 15 minutes). Again though, a handful of interviewees reported that their average teletext session lasted from one-half hour to one-and-one-half hours. Certainly, larger surveys will be needed to confirm or deny these figures as representing national averages.

Both the BBC and ITV began to promote actively their teletext services in the early 1980s to a much greater extent than during the previous six years. The BBC's primary procedure for promotion was to begin what they called "In-Vision." In this service, teletext pages are broadcast as normal video programming during those parts of the day when the BBC is not usually on the air. In this way, viewers with any type of television set can see Ceefax pages, although the pages roll by without any selectivity on the part of the viewer. (This technique is also being used in Chicago on WFLD–TV, which is described in the next chapter.) The British government also took a hand in promoting teletext and helped to finance a National Teletext Month in October 1981.

At the same time, ITV launched a $5 million campaign to advertise Oracle. This was ITV's first major effort to promote their teletext service and was used to announce the arrival of paid teletext advertising on Oracle itself. The success, or lack of it, of advertising on teletext pages will probably be a significant factor affecting teletext in the United States, and the Oracle experience is being watched closely. The Oracle rate card for advertisements, published in late 1981, provided for two classes of ads—a full page for about $800 per week, and a fractional page (two lines of text on an existing page) for $600 per week if the sponsor selects the page or $400 per week if Oracle selects the page. In addition, Oracle contains an index to the advertisement pages, and short announcements on other pages alert viewers to the index. Because teletext pages can be changed very easily, Oracle offered to change the text of any ad during its run for an additional $20 per page or partial page.

Besides being the lead service in teletext advertising, Oracle also took the lead in regional teletext. The concept of regional teletext means that stations in a geographic area can insert their own teletext pages into the nationally distributed magazine. This was first accomplished in late 1981 as Scottish Television, one of the ITV stations, prepared to originate about 60 pages of local information.

In general then, British teletext provided the early model for teletext services, and British equipment was often used for teletext systems in other countries. The Ceefax and Oracle services provide news, weather and sports information, games, puzzles, recipes, horse racing results, fictional stories broadcast in installments over a number of days, personalized messages, and special interest group information such as prices and crop disease warnings for farmers.

The British systems have also continued to grow in capacity and capability. The BBC's initial service used an Alpha LSI–2 minicomputer. This was replaced in 1979 by a set of three PDP 11/34 minicomputers from Digital Equipment Corporation, with capacity for 10,000 pages of text and graphics. (Of course, a teletext magazine at any one time still contains only about 100 pages.) The system also has several automatic links to news services, the London Stock Exchange, and the national weather service, so that information can be fed into the teletext system without being retyped. Even color radar maps are automatically transferred from the weather service to Ceefax.

Other enhancements that have been added include fine-line graphics, full-color photographic-quality pictures, and telesoftware. The latter term refers to the use of the teletext pages to broadcast computer programs and data to television sets equipped with a microprocessor, or to a microcomputer with a teletext decoder. In the late 1970s, Oracle engineers began to test the technical feasibility of telesoftware in the belief that this could be a way to support nonuser-programmable microcomputers in homes [7]. People could have the power, and fun, of using microcomputers without knowing a thing about programming. The programs for games, educational courses, financial analysis, and the like would be broadcast using the vertical blanking interval, and the receiver would be a smart but familiar television set. Whether or not telesoftware develops in that way remains to be seen, because the basic concept can be accomplished with cartridges, or with programs downloaded over a cable television system, as is the case with Playcable and Mattel's Intellivision in the United States (see Chapter 4).

The transmission of full-color "photographs" via the vertical blanking interval has also been demonstrated, with the expectation that eventually the U.K. teletext systems will begin transmitting both single photographs and sequences of photographs, as a sort of slow-scan digital television [8]. At present, however, the transmission of a full-screen photographic picture using teletext is equivalent to broadcasting about one hundred "normal" teletext pages. In addition, other enhancements call for the transmission of separate data "channels" unassociated with the display portion of the teletext page.

Several other developments involve the use of subcodes to identify each page. In the current U.K. teletext system there is room for several million subcodes. These can be used to provide unique addresses, so that certain coded pages can be seen only by a person, or group of persons, with a decoder that has the proper address built in. Or the subcodes can be used to digitally classify the pages so that a viewer can choose which pages to see by selecting a category, rather than an individual page.

The current generation of teletext sets in Great Britain contain an internal digital memory of one thousand bytes, enough to hold a single screen or page of text. But as the cost of semiconductor technology declines, additional memory can be built into the set. This will allow the viewer to locally store a set of pages, which could drastically reduce the waiting time necessitated by the broadcast nature of teletext. In fact, a memory of, say, one million bytes could hold a thousand pages, perhaps automatically selected and stored using commands preset by the user.

The use of Ceefax and Oracle for subtitling, or closed captioning, has also continued to develop. Here the problem is not really a technical one related to teletext—the system handles several streams of subtitles easily—but a greater problem related to converting speech to print in real time. The television programs recorded prior to transmission can certainly be subtitled at a much slower pace, with subtitles keyed to individual frames and coded for color and location on the screen. But the captioning of live transmissions requires some extremely fast procedure for converting sound to accurately printed words. The BBC has experimented with typists using a stenographic machine linked to a computer, such that the computer can search its memory for the correct spelling of the word and phrase symbols generated by the stenographic typists. The result has not always been as desired; a live telecast of a speech by U.S. President Ronald Reagan carried in the captions continued reference to "free dumb " (freedom). At ITV, Oracle has reportedly captioned some live programs simply by hiring an extremely fast typist who managed to get most of the spoken words into the captions.

To a great extent, the development of teletext in England spurred the introduction of teletext services elsewhere that could begin with some of the features that started out as enhancements to the U.K. systems. Almost as soon as the British system became a public service, engineers in other countries began developing better systems, that is, systems that would overcome minor problems with the first versions of the U.K. system. It has been said that by the end of 1978, representatives and interested parties from over one hundred countries had visited the BBC and ITV to see just what teletext was all about. Some of these researchers returned to their respective countries to apply what they had learned to change existing information delivery systems or to create new systems. For example, one of the simple but important features of the U.K. system that others admired was that, in one way, the system provided an immediate response to a viewer's command even though the viewer might still have to wait for a page to appear. This was accomplished by reserving a portion of the top line of each page for a page-number display. When a viewer selects a page, the top line display immediately changes color and begins showing in rapid succession the numbers of the pages being broadcast. The viewer can see the numbers flipping by and know fairly well when the desired page will appear; this has the effect of reducing the perceived waiting time for page access.

Part of the rise of competitive systems elsewhere, then, came from this ability to learn from the English pioneers, but another part is also due to the fact that television technical standards differ throughout the world, and the English system was originally designed for one particular standard to accommodate one particular written language. Thus there was a lot of debate, for example, on how well the British teletext system based on a television standard of 625 scan lines and 50 Hz operating power could be converted to, say, the U.S. system of 525 scan lines and 60 Hz operating power. The conversion was certainly possible, as later chapters will describe, but some changes did have to be made. And in countries with different, and larger, character sets for written language, still more changes had to be made. On the other hand, in countries with the same television standard as Great Britain's, and similar character sets, the adoption of U.K. teletext methods has been relatively simple. By 1981, as teletext began to grow substantially in Great Britain, the U.K. form of teletext was being used in about a dozen other countries.

Japan

In Japan, the development of teletext goes back about as far as it does in Great Britain. But the Japanese effort did not result in the early introduction of a public service. Instead, teletext is only one part of a wider series of technical and social experiments with new information delivery systems, including such means as videotex via fiber optic cables.

The first teletext system in Japan was probably developed by Matsushita Electrical Industrial Company in early 1969. Like the early British efforts, this was an attempt to create a system of broadcasting digital information to printing devices. And also like the British experience, the major broadcasting organization was involved. Matsushita was working under the direction of the Japan Broadcasting Corporation (NHK—Nippon Hoso Kyokai). By 1971, several years before the British had brought the Ceefax/Oracle system and the Prestel system (which uses telephone lines) together under one standard, the Japanese were developing a television set that would receive both broadcast teletext and telephone-delivered videotex.

In 1976, the first formal technical guidelines for teletext were proposed by NHK, and in 1978 experimental teletext broadcasts were begun (see Figure 2.2). Teletext has continued on an experimental basis, with preparations for a commercial service to begin in late 1983.

But even though the Japanese have proceeded rather cautiously in approaching a teletext service, they have been centrally involved in technical teletext work in other countries. Japanese engineers from Sony Corporation took part in the 1978 CBS teletext tests, and Sony has supplied teletext-equipped television sets for Great Britain.

One of the major characteristics of Japanese teletext and videotex systems, and the

Figure 2.2. Sample NHK teletext page. *(Courtesy of Japan Broadcasting Corporation.)*

reason for the long period of development, is the need to accommodate the complex character sets used in Japan. The characters are not only complex but also numerous. There can be as many as 3,000 characters in the Kanji character set, for example. Therefore the Japanese designers had to construct a teletext system that could display that many characters, as well as Katakana and English characters, and maintain the legibility of the characters on a television screen.

The result was a teletext specification that calls for 15 characters in a row and only 8 rows on a screen for Kanji characters. Because it was not economically practical to build a character generator into the television set for thousands of characters, the patterns of dots to create the characters are broadcast instead. The television screen thus becomes a matrix of dots—248 dots wide and 200 dots high. This permits fine-line graphics in addition to the complex characters. The memory required in the television set to hold the patterns of dots is considerable, though. Large enough for approximately fifty thousand bytes, it is fifty times the size of the memory in the current British teletext decoders.

Similar to the U.K. teletext standard adapted for the United States (the United States and Japan have the same technical standard for television), Japanese teletext operates at a data rate of 5.72 megabits per second and has the same structure for organizing the digital information on television scan line. There are a few differences, however, such as the way colors and other attributes are coded and transmitted, and the way scrolling is handled. Japanese teletext also comes in several versions; one of the versions provides several lines of text moving horizontally across the bottom of the screen superimposed over the normal television program and selectable by the teletext viewer.

Besides the teletext activity, Japan has fostered a variety of videotex experiments. One system uses coaxial cable, the Coaxial Cable Information System (CCIS), and has been tested in a part of Tokyo called Tama New Town. The CCIS trial service, begun by the Ministry of Posts and Telecommunications in 1976, has experimented with facsimile newspapers, still pictures on demand, automatic news alerts, and hard copy dot printers in the home. Another trial service, Captain, which began in 1979, is a telephone-based system developed by Nippon Telegraph and Telephone. Along with the usual features of a videotex system, Captain has an audio facility so that both images and sound can be delivered to a user's terminal. A third trial service, called HI–OVIS (Highly Interactive Optical Video Information System), is based on the use of fiber optic cables as the distribution medium.

The Japanese influence on worldwide teletext development has yet to be strongly felt. Initially, Japanese teletext (and videotex) was considered a ''local'' development in that the system had to function with so many character sets and consequently was unlike the other systems in the United States and Western Europe. However, there are several reasons to believe that the Japanese system may have a lasting if belated effect throughout the teletext world. First, the Japanese system incorporates many features deemed highly desirable, such as fine-line graphics and scrolling text. Second, Japan is in the midst of a massive effort to lead the world in developing very large-scale integrated circuits, which are the building blocks of future teletext systems.

France

Teletext history in France goes back to about 1973, when the U.K. teletext system was demonstrated in France and French researchers subsequently conducted a feasibility study to see if the U.K. system could be adopted without change. But like Japan, France had trouble with character sets. The French language has more characters (i.e., characters with diacritics) than the English language, and the early British teletext system did not incorporate these other characters. The result, as in Japan, was the creation of a teletext system to meet domestic requirements.

One of the often mentioned aspects of French teletext is that it uses the same packet structure for digital information whether the packet is sent over a telephone line or broadcast within a television signal. The reason for this is often given in terms of compatibility by design, but the multiuse packet structure may equally be a result of the fact that the initial teletext studies in France were entrusted to the same group that was designing a national packet network for data communications. The CCETT (Centre Common d'Études de Télévision et Télécommunications), owned jointly by the French telephone authority and the broadcast authority, was given the task of proposing a teletext system suitable for France at the same time that they were designing the national packet network [9]. Because the CCETT staff was rather small at the time, it was easy to share the basic packet structure concepts.

The proposal, presented in 1976, described both a display coding system and a transmission coding system. The display scheme, called Antiope (Acquisition Numérique et Télévisualisation d'Images Organisées en Pages d'Écriture), at the time offered improvements in display over the existing British system, primarily in the ability to display different colors and other attributes without leaving blank spaces on the screen. The transmission coding scheme is called Didon (diffusion de Données) and is the data packet structure for transmission via any medium, whether by telephone line, cable, or broadcast television signal. The advantage of the Didon/Antiope combination was this employment of a ''universal'' packet structure. The disadvantage in comparison with the U.K. form of teletext was that a more complex teletext decoder was required because the Didon packets had to be interpreted and the Antiope codes processed before a display could be produced.

Télédiffusion de France, after technical tests of the proposed teletext system, began a teletext service in 1977 aimed not at the general public but rather at a limited market segment—the financial community. The service is called the Antiope-Bourse (the stock exchange service), and broadcasts teletext stock reports on over 2,000 stocks to brokers and money managers in Paris and Lyon. However, Antiope-Bourse, which became a public service in 1979, does not merely use the vertical blanking interval of an existing television channel; instead, it has its own channel with no associated video programming, so that many scan lines are available for carrying digital signals. The goal is eventually to use the full channel for multiplexing the digital information (i.e., full-channel teletext). The channel being used belongs to the old 819-line network that was being phased out in France and thus was available for full-channel broadcast teletext. In early 1981, the Bourse service was using 60 scan lines for data, with the intention of eventually expanding to about 300 scan lines. The latter number of lines will be able to carry between 5,000 and 10,000 pages of broadcast teletext with reasonable average waiting periods between page selection and page appearance.

The full-channel teletext system as developed in France still has the familiar structure of magazines and pages, with about 100 pages per magazine. Within that structure, several transmission schemes have been designed to give a user the most rapid page access possible given the nature of broadcast teletext [10]. One procedure is to assign a certain number of scan lines to each magazine, or more generally, to a digital subchannel. A user would then select the subchannel or desired magazine before requesting a page. An alternative procedure, which would require more processing power, would be to dynamically allocate the scan lines as pages are ready to be broadcast.

Aside from the fact that Antiope-Bourse is a specialized service that uses full-channel techniques, there is another notable difference between the beginnings of teletext in France and the beginnings in England. In England, the BBC inaugurated Ceefax with teletext editors virtually on-line, typing the data into the system and sending out news updates as fast as the news came in over the wire services. In France, however, the initial effort concentrated on an automatic link between the stock exchange computers and the teletext computer. The teletext system merely broadcast the information flowing in from the stock market computer system.

The philosophy of structuring various teletext services for special audiences is behind the other teletext trials in France as well. The Antiope-Postes service was designed to broadcast a teletext magazine of interest to post office employees. The Antiope-Meteo, or weather service, was aimed at workers in the transportation, tourism, and agricultural industries. And Antiope-Orep was developed for a specific geographic area of France, the southwest. These services have yet to experience the growth in teletext viewers that took place in England in 1978–1981. As of mid-1982, the estimated number of teletext-equipped television sets in France totaled fewer than two thousand.

Similar to the situation in Japan, teletext in France is only one part of a many-pronged effort in computer/telecommunications to prepare for the information society and the information economy. In 1975, the French government began a massive drive to modernize the cumbersome and ineffective telephone system as part of a larger effort to install the most modern data communications facilities and computer-based communications services. This has become known as the Telematique Program. The Telematique umbrella covers not only teletext and videotex, but also electronic telephone directories, a photocopier/facsimile machine for the mass market, an electronic blackboard for telewriting, a version of teletext that would allow a multiplexed code to turn on individual video tape recorders, and a "smart card" or credit card with a built-in microprocessor. All these products, services, or techniques are being developed as part of the centrally directed Telematique research and development program.

In short, while England was building a national up-to-the-minute teletext news and information service for the mass market, France concentrated on specialized teletext services. But more important than that, France spent almost as much effort in selling Antiope abroad, specifically in the United States. The international selling effort, called Intelmatique, functions as the official representative of the French Direction Générale des Télécommunications. One of the early results of this effort was the adoption by the CBS television network of a modified Antiope system for teletext in the United States, which is covered further in the next chapter. Whether French teletext will blossom on its own or only follow a successful Antiope service elsewhere remains to be seen.

West Germany

In the early 1970s, paralleling the developments in a number of other countries, West German television organizations were investigating ways of broadcasting subtitles on a selective basis to television sets with the necessary decoders. In 1975, however, British teletext engineers assisted the Bavarian Wendlestein television station near Munich in broadcasting experimental teletext signals, and the previous subtitling approaches were dropped in favor of using the more inclusive teletext [11]. By early 1976, a government report concluded that teletext services were both technically and economically attractive in West Germany.

The next few years were occupied with technical testing and decidedly nontechnical political arguments. While some political groups were worried about the broader social implications, the broadcasting and publishing industries were more concerned with the more immediate debate over who should develop teletext—the broadcasters or the publishers. On the technical side, the tests were accomplished by broadcasting a test page and measuring reception at over 1,300 sites around the country using mobile test equipment. The results indicated that although only 72 percent of the measurement sites could receive normal video at a ''fair'' level or better, 77 percent of the sites could receive error-free teletext using the British system. Two notes that can be made about the tests are that ''fair or better'' reception is a subjective evaluation, and 9 percent of the sites that did receive fair or better video pictures were unable to receive the teletext test page error-free. In a related test at the antenna sockets of about 270 homes, the overall success rate of teletext reception was 90 percent, and if only those homes where normal television reception was fair or better were counted, the teletext success rate was 94 percent. [12].

Teletext services were also demonstrated during this time, principally at international radio and television fairs in Berlin. The exhibitions' teletext services showed subtitling in various languages, television program information, question and answer pages, and schedules of events. In 1977, the teletext signals were broadcast only in the Berlin area, but during the 1979 fair, the teletext pages were broadcast throughout West Germany via the national television networks. In 1980, the German authorities began also testing a French Antiope system in order to compare it with the British system in use.

By the early 1980s, West German television had begun an initial teletext service based on British technology. However, the result of the political discussions of the several previous years was that some restrictions were imposed on the teletext service, governing when teletext could be broadcast during the day and prohibiting the teletext service from broadcasting information that had not yet appeared on the normal television channel.

As in France, the West Germans had difficulties at first with the British system because of the need to display more characters than are used in English. Two different methods of utilizing the British system to accommodate the additional characters were subsequently developed and studied. The method eventually chosen was one that had originally been developed in the Netherlands for the same purpose.

As in Great Britain, Japan, France, and other countries, teletext trials in West Germany paralleled videotex ventures. (The German videotex system is known as Bildschirmtext.) But West Germany has played more of a coordinating role than a creative one. It did not set out to develop its own technical system for teletext after looking at the other entries. Instead, the available systems were evaluated and modified as necessary. In the position of mediator between rival systems, West Germany has gained credit for helping to bring about a European videotex/teletext standard that embraces the British system, the French system, and the West German enhancements and is supported by 26 European countries (see Chapter 5).

Canada

Canada is the third member of the triumvirate of Britain, France, and Canada that dominated teletext talks in the late 1970s. It was widely believed that there were only three real teletext systems, from these three countries, and that these systems were rivals—and not compatible. At the time, the systems were indeed not compatible and did not have the same features, but that situation did not last. The degree of compatibility and the rise of other systems, however, is the subject of later chapters.

Canadian teletext is more or less an offshoot of Canadian videotex, which was created as a direct response to the British videotex system, Prestel. This is not to say that Canada ignored videotex and teletext prior to Prestel. On the contrary, Canadian researchers, like researchers in most other industrialized countries, were investigating the potential of the computer/telecommunications merger for mass market information services. In addition, during that period the Communications Research Center of the Department of Communications was assisting the Canadian satellite program with work on computer graphics and modeling techniques [13]. By 1976, the Communications Research Center had pretty much completed work on a graphics system using microprocessors as the display terminals and utilizing graphic drawing commands as the basic structure for coding pictures. As an example, a circle would be drawn at a display terminal when the circle command was transmitted to the terminal along with positioning and size information. Virtually any kind of picture can be drawn using just a small number of graphic commands.

This approach was standard computer graphics design but quite unlike the mosaic-style graphics developed in England and France, and also unlike the bit-pattern approach used in Japan, for teletext and videotex. Therefore when this system evolved into Telidon, the Canadian teletext/videotex system, it was perceived as a new and advanced way of handling teletext and videotex graphics. And unlike the earlier systems, it was obvious that Telidon was heavily graphics-oriented (see Figure 2.3).

The actual metamorphosis of the Research Center's graphics system into Telidon has been attributed to a spirit of competition with the foreign systems. In 1977 and 1978, it looked as if Canadian companies were going to begin using the British Prestel system and, to a lesser extent, the French Antiope system. Meanwhile, the Communications Research Center had created its own Prestel-like system but was convinced that the graphics-command approach produced a much better system. More important, the Department of Communications was able to convince the government that the graphics-

Figure 2.3. Sample Telidon page. *(Courtesy of Infomart.)*

command system was indeed better (and a native system at that); the Canadian government then agreed to fund a program to spur the introduction of Telidon into the marketplace.

Teletext itself got underway in Canada in early 1980, although the initial studies began as early as 1976 and were based on the British teletext system [14]. The Ontario Educational Communications Authority (OECA), with nine television transmitters, was interested in applying the teletext technique to the delivery of educational material. One of the basic problems with the U.K. system, though, at the time that the OECA wanted to begin teletext was that it had not yet been adapted to 525-line NTSC television (the technical television standard used in Canada, the United States, and two dozen other countries). When the French Antiope system was announced, it seemed more readily adaptable to North American television, and the OECA began preparations for a trial service, based on Antiope, in 1978. But the planning was temporarily halted when Telidon was proclaimed the Canadian answer to teletext and videotex. The decision to use Telidon soon followed, putting off the start of the teletext trial until 1980 because plans had to be reformulated and equipment secured.

The overall result of the replanning was that OECA began a four-level trial in 1980: a broadcast teletext service; an interactive telephone line-based service; participation in Bell Canada's own separate videotex trial; and participation in several cable television-based Telidon trials. The OECA's teletext trial ended its first phase in early 1982 and began preparing for a second two-year phase starting in late 1982. During the trial, a total of 55 teletext-equipped television sets will be rotated among schools, libraries, museums, and a few homes. The teletext signals are broadcast in the vertical blanking interval of TV Ontario, the educational television network in Ontario.

The teletext magazine broadcast by TV Ontario, called Edutel, consists of both educational and general purpose information, although the OECA has stated that it will broadcast only educational teletext once other television stations begin teletext. The educational portion of the teletext magazine includes not only learning programs but also program notes, teacher's guides, course listings at various schools, and news about jobs in education, school bus status, and other administrative matters.

The OECA teletext trial involves technical testing as well as testing the design of services. At the time the trial began, OECA and the Department of Communications were still unsure about the best data rates to use for broadcast Telidon and the best way to integrate the teletext signals with the test and reference signals already in the vertical blanking interval. And because the OECA broadcasts to remote areas using a direct broadcast satellite, teletext could be tested in that environment also.

At the conclusion of the trial, the OECA hopes to learn some relatively minor details, such as the maximum acceptable waiting time for a teletext page to appear, and some major evaluations, such as the educational effectiveness of the learning material distributed via teletext and the social impact of using such systems. And of course the technical reliability of the system will be assessed.

In mid-1981, the Canadian Broadcasting Corporation announced that it too would begin broadcasting teletext using Telidon, with the service to be on the air in 1982 or 1983. The service is to be in both English and French with separate teletext magazines for each and will be distributed nationally by satellite. The teletext-equipped television sets will be placed in public places across the country and in some private homes. A total of 750 sets will be moved periodically until 1,400 homes, selected by CBC Research, have had a chance to view teletext.

The types of teletext information to be broadcast will be a mix of local, regional, and national items. Two likely first applications within that breakdown are election-related information and grain and livestock prices for the agribusiness community.

The technical considerations for the CBC's trial are substantial, because regional production centers will need computer power to manipulate incoming teletext and to insert locally originated pages. Moreover, delay facilities at six of the production centers will need to be bypassed. Technical control will be accomplished with local processors at the production centers communicating with a central computer that will serve as both the host or storehouse of the teletext pages and as the transmission traffic cop. The CBC's teletext trial will run until 1985.

In addition to the teletext trials in Canada, videotex trials of Telidon have sprung up in virtually every province (see Chapter 6). Beyond that, Telidon has made its mark in the United States by being chosen for a publicly funded teletext trial (see Chapter 3) and for several cable television trials (see Chapter 4), and by being adopted and adapted by AT&T as part of the AT&T "standard" for videotex and teletext display that in turn is to be adopted by the American National Standards Institute (see Chapter 5).

The Canadian entry into teletext was somewhat unlike the experience in France or Japan, where the initial impetus came from a need to accommodate local language characteristics. Instead, Telidon was produced as a direct competitor because of the belief that it handled some things (e.g., graphics) much better than any other system. As will become evident, that only prompted the other systems to develop similar features, but the fact remains that it was the Canadian influence, coming out of the computer graphics world, that instigated the first major improvement to the growing realm of teletext (and videotex) systems.

Elsewhere

At least a dozen other countries have witnessed the start of broadcast teletext, with most of the teletext-equipped sets in use to date containing the British technology. In Western Europe and even Eastern Europe, Australia, and the Far East, teletext services are taking hold.

Sweden

The Swedish Broadcasting Corporation began teletext broadcasts in 1979 primarily as a service to the hearing-impaired. Using the British system and old BBC equipment, the service, called Extratext, was begun; a group of 160 people with hearing losses was selected as the audience, and several hundred pages of teletext were broadcast. During 1980, the Swedish Broadcasting Corporation developed its own computer system for the service and began expanded teletext broadcasts. In mid-1982, there were an estimated 110,000 teletext sets in use, with that number growing by some 2,000 sets per month.

The future for teletext in Sweden looks promising. Industry projections are that by 1985 all television sets of a certain size being manufactured will have an integral teletext decoder [15].

One of the aspects of Swedish teletext that is often mentioned is the effect of government planning in the whole area of computerized communication and information services. The Commission on New Information Technology—one of a long line of similar commissions—was established in 1978 to present recommendations to Parliament covering such items as government use of teletext, public access, and rules governing content, advertising, copyright, and funding of the teletext service. The recommendations were presented in 1980, and it is expected that they will guide teletext development, although the Swedish Broadcasting Corporation had already published its own decisions regarding the future of teletext.

The Netherlands

Also using the British system, the NOS (Netherlands Broadcasting Corporation) television network in the Netherlands has been broadcasting teletext since 1979, beginning with a one-year trial service. In January 1981, the teletext service, called Teletekst, was established on an expanded trial basis and in 1982 extended for another two-year period of "experimentation" (see Figure 2.4). The service is financed in part by the Dutch

Figure 2.4. Sample NOS teletext page. *(Courtesy of Netherlands Broadcasting Corporation.)*

federal department of broadcast affairs. The fact that there are over 150,000 teletext-equipped television sets in the Netherlands seems to assure the future of the serivce. And the size of the teletext audience has already sparked some user surveys. One survey by NOS, the national association of broadcasters, concluded that the most popular pages were news and consumer information and the weather map. Although the survey did not reveal daily viewing habits, most respondents to the survey said that they watched at least once a week, in the early evening.

Austria

The Austrian television authority, Österreichischer Rundfunk (ORF), began a small teletext service in 1980 using a system built according to the British standard. The first service contained only about 64 pages of teletext information. In July 1980, however, ORF purchased a much larger teletext system almost identical to the one in use at the BBC but with a few additional features such as improved handling of subtitle creation. ORF also established a teletext newsroom much like the one at Ceefax. At the time of the installation, there were an estimated 15,000 teletext sets already in use in Austria (see Figure 2.5, showing the use of teletext for closed captioning).

Besides the normal teletext broadcasts, ORF broadcasts the teletext pages as normal video during certain periods of the day. This is the same technique used by the BBC and by WFLD–TV in Chicago to popularize teletext by showing *all* viewers what the teletext pages look like, even though selectivity on the part of the viewer is impossible. These ORF teletext pages are also viewable in Hungary, where a Hungarian teletext service is being created.

Belgium

Two different networks in Belgium, BRT and RTBF, decided to use two different technologies to begin broadcasting their respective teletext services. The BRT network got on the air first using the British teletext technology in 1980, followed by RTBF in 1981 using the French Antiope system. Because BRT caters to the Flemish-speaking com-

Figure 2.5. Sample ORF teletext page. *(Courtesy of Österreichischer Rundfunk.)*

munity and RTBF serves the French-speaking population, the technological split matches a cultural split. It is unclear whether one of the technologies will eventually predominate, but it is likely that the two systems will coexist for some time.

Switzerland

Swiss television began teletext broadcasts on a trial basis in 1981. According to one report, the equipment used is the old British-based system first used in Austria before ORF installed the newer three-computer system.

Denmark

In Denmark, the first teletext transmissions actually began in 1977, using the British teletext system. But since then development has been quite slow because of disagreements among various interest groups, led by newspaper and magazine publishers.

Hungary

Hungarian research into teletext began as soon as the U.K. specifications for teletext were published. The first studies were carried out at Budapest Technical University and subsequently, in 1978, the university built a small teletext system using a microprocessor. The page capacity was minimal, which did not really matter, because the system was used only to produce test pages for field trials that began in 1980. Because the technology under trial was British, the university researchers decided to test the French system as well, beginning in 1981. Following the conclusion of the tests, an experimental service is to begin in 1982 or 1983.

Hungary is a good example of the difficulties caused by differing technical standards. Because television sets in Hungary can receive broadcasts from stations using the PAL standard for color television and from stations using the SECAM standard, all color sets in Hungary must now be able to operate on both standards. This naturally makes the sets more expensive, and Hungarians would not like to see the same thing happen with regard to teletext. Hungary has been somewhat spared the problems associated with different character sets by virtue of adopting a Swedish-built teletext system, as the Swedish alphabet contains almost all the diacritics used in Hungarian.

Australia

Over a half dozen television stations throughout Australia broadcast teletext. Stations in Sydney, Melbourne, Adelaide, Brisbane, Newcastle, and Perth have all put teletext magazines on the air, some of them beginning in 1979 and others in 1980. The Australian Broadcasting Commission has also tested teletext on its own stations and switched in 1981 from a British system to the French system to try to solve some reception problems that had plagued the tests.

Australia may have the distinction of being the first country where a television station claimed to be making a profit on its teletext service. Brisbane TV began selling three-line advertisements on most of its pages and by late 1980 stated that advertising revenue was covering personnel costs (two editors), operational costs, and equipment costs. At the time, an estimated 1,500 families in the Brisbane area had teletext-equipped television sets. Nationally, the total number of sets in use was somewhat over 10,000 in 1982.

Table 2.1. Estimated Worldwide Population of Teletext-Television Sets

	Number of Sets		
Country	1974–1980	1981	Mid-1982
United Kingdom	100,000	300,000	500,000
West Germany		90,000	150,000
The Netherlands		10,000	150,000
Sweden		90,000	110,000
Austria		20,000	105,000
Belgium		10,000	20,000
Finland			20,000
Australia		10,000	12,000
Switzerland			10,000
France		1,000	2,000
United States		200[a]	300[a]
Denmark			300
Norway			300
Canada		50	60
Yugoslavia			NA
Israel			NA
Hungary			NA

[a]ANA—not available.
Figure does not include some 50,000 sets able to receive the line 21 form of teletext.

The extent of teletext activity is considerable. The addition of a simple teletext service to an existing broadcast television signal can be a rather inexpensive proposition. Even the most advanced computerized teletext system can cost less than $300,000. The main problem, though, is not in originating the service but in seeing it. This requires a turnover in the television set population, as sets with integral teletext decoders become less expensive and readily available, or a less expensive set-top adapter is produced. (The set-up adapter, however, of necessity produces a worse picture because the teletext page must be modulated to pass through the antenna connectors, while in a teletext set with an internal decoder the color information for the created page can be sent directly to the color guns.) As Table 2.1 shows, the worldwide population of teletext sets in 1981 was estimated to be over a half million, and in mid-1982 over a million, but that is barely the beginning. One of the largest markets, the United States, has yet to weigh in, although here too considerable activity has taken place, as described in the following chapters.

References

1. Morgan, Gwyn, Inside Teletext, *Hobby Electronics,* September 1980, pp. 12–16. See also Gwyn Morgan, Teletext—Present and Future, *Communications Engineering International,* February 1981, pp. 19, 21–25; Colin McIntyre, Teletext in Britain: The Ceefax Story, in *Videotext,* Efrem Sigel (ed.), Knowledge Industry Publications, White Plains, N.Y., 1980, pp. 26–48.
2. Sherry, L. A., Teletext Field Trials in the United Kingdom, *IEEE Transactions on Consumer Electronics,* CE–25 (3): 409 (July 1979).
3. McIntyre, Colin, Teletext in Britain: The Ceefax Story. In *Videotext,* Efrem Sigel (ed.), Knowledge Industry Publications, White Plains, N.Y., 1980, p. 29.

4. Philips Video, "Teletext and the Consumer," July 1981, p. 8.
5. British Study Identifies Favored Teletext Users; Confirms Upscale Nature, *Videotex Teletext News,* November 1981, pp. 9–10. See also Philips Video, "Teletext and the Consumer," July 1981.
6. Herring, William, "Teletext User Survey. Ceefax and Oracle—User Reactions," 1982, distributed by the Videotex Industry Association, Washington, D.C.
7. Hedger, J., Raggett, M., and Warburton, A., Telesoftware—Value Added Teletext, *IEEE Transactions on Consumer Electronics* CE–26 (3): 555–566 (August 1980).
8. Chambers, John P., Enhanced U.K. Teletext Moves Towards Still Pictures, *IEEE Transactions on Consumer Electronics* CE–26 (3): 527–533 (August 1981). See also R. H. Vivian, "Level 4 Enhanced UK Teletext Transmits Graphics Through Efficient Alpha-Geometric Coding," Independent Broadcasting Authority, 1982.
9. See, for example, B. Marti, Broadcast Text Information in France, *Viewdata 80* (Northwood Hills, England: Online Conferences Ltd., 1980), pp. 359–369.
10. Berger, M., Diode: A Full-Field Teletext System, *Viewdata 81* (Northwood Hills, England: Online Conferences Ltd., 1981), pp. 25–31.
11. See, for example, Ulrich Messerschmid, Teletext in the Federal Republic of Germany, *Viewdata 80* (Northwood Hills, England: Online Conferences, Ltd., 1980), pp. 431–445.
12. Ibid.
13. For a good introduction, see Herbert G. Bown and William Sawchuck, Telidon—A Review, *IEEE Communications Magazine,* January 1981, pp. 22–28.
14. Bowers, Peter G., and Cioni, Maria, Telidon and Education in Canada, *Viewdata 80* (Northwood Hills, England: Online Conferences, Ltd., 1980), pp. 7–17.
15. Ohlin, T., Videotex and Teletext in Sweden—A Nation Decides, *Viewdata 81* (Northwood Hills, England: Online Conferences Ltd., 1981), pp. 215–230.

3

Broadcast Versions in the States

In 1970 or 1971, broadcast engineers in the United States were investigating ways of using the vertical blanking interval to carry information. Some of the possible uses were strictly for test and measurement purposes. But beyond that, as in Europe and Japan, it was recognized that the vertical blanking interval could be used to enhance television for people with hearing disabilities. Digital information representing text could be multiplexed within the television signal and selectively received, with no negative effects on the television signal itself.

One of the earlier proposed uses of the vertical blanking interval—to carry time and frequency data—was suggested by the National Bureau of Standards in 1971. The ABC television network then suggested that the same system be used for closed captioning and subsequently demonstrated, in conjunction with the National Bureau of Standards, this use of the vertical blanking interval in late 1971 and early 1972. In January 1972 a committee of engineers led by members of the ABC technical department was formally established to study this form of teletext. The committee originally considered using line 1, but tests by RCA showed that this would cause interference with the broadcast program. The Public Broadcasting Service (PBS), which by this time had joined the effort, then suggested using line 21 after the committee had considered lines 19 and 20 as an alternative to line 1. Thus the so-called line 21 system began.

During the next two years, 1973–1975, PBS conducted a series of field tests examining the National Bureau of Standard's scheme for inserting data into a television signal, as well as testing an alternate scheme developed by Hazeltine Research, Inc. [1].

During that time, all PBS stations broadcast imbedded test data at one time or another, and 20 of the stations using prototype receivers tested the viewer reaction of people with hearing impairments.

This activity culminated in a request to the FCC in 1975 for a rule-making establishing permanent use of line 21 (in field one and half of field two) for closed captioning. (Closed captions are captions that can be seen only by viewers with the proper decoder.) Not surprisingly, the PBS plan was not welcomed by everybody, and the CBS television network, the Electronic Industries Association, and the National Association of Broadcasters all told the FCC that further studies should be made, especially of the use of the teletext systems then emerging in Europe as vehicles for closed captioning. During the next four years PBS persisted, and the National Captioning Institute was eventually created to actually perform the captioning tasks. Broadcasts of programs using the line 21 system for closed captioning did not begin on any sort of scale until 1980.

This long gestation for teletext (or rather the closed-caption version of the teletext technique) has been attributed to a number of factors. A primary cause was said to be the regulatory environment. The FCC would try to win some form of consensus before proclaiming a technical standard, and it can be subject to heavy political pressure. Other suggested causes are the size of the U.S. market (making the task of obtaining industry-wide agreement on a plan of action exceedingly difficult) and the commercial nature of U.S. television. The latter fact was certainly behind comments by industry executives in the early 1970s that teletext would have considerable difficulty in the United States because broadcasters would not willingly give viewers something else to turn to when commercials came on. It apparently never occurred to these commentators that teletext could carry advertising too.

All these reasons are valid, but they do not explain the entire situation. At least three other circumstances should be noted. The first is cable television. In the early 1970s, cable television was billed as a cornucopia of new services, and much of the information-providing nature of teletext was being talked about and tried as a function of cable television. This, and the later adoption by cable television of the teletext technique, justifies placing the cable teletext story apart from broadcast teletext (see Chapter 4). The second reason behind the slow growth of teletext is that some of its more talked about capabilities, particularly in England, such as the ability to provide newsflashes easily, were also being pursued in the United States but in a different way. Television stations began using simple character generators to send messages—or newsflashes—crawling across the screen when a newsworthy event warranted the intrusion. Finally, one of the driving forces behind the development of teletext, closed captioning for the hearing-impaired, was also being answered in a slightly different way, namely by being attacked as the primary problem to solve. Moreover, the solution encompassed more than the technical specifications for the line 21 system. Along with the technical system, the National Captioning Institute was founded so that there would be captioned programs available when the viewing devices, the decoders attached to television sets, became available. This is opposite to the situation in England, where the technical system was established and teletext went on the air long before any substantial effort was made actually to prepare captions for programs.

This chapter tells the story of the long slow start of broadcast teletext in the United States. The line 21 system and several special purpose teletext systems were created and implemented during the latter half of the 1970s, and only as the 1980s started did a few "mass market" teletext trials get underway. As these latter trials unfolded with their relative handful of viewers, the line 21 system showed its flexibility for use as a text service, and other special teletext systems such as subscription teletext showed signs of pointing the way for teletext in this country, indicating that teletext might develop as a special service or limited market technology before the mass market applications would be able to grow.

Early Entrants

About the time that PBS was petitioning the FCC to protect the use of line 21 for closed captioning, another form of teletext was being tested in Philadelphia—not by a ''normal'' television station but by a special broadcaster, an MDS station. (Multipoint Distribution Service stations use super high frequencies to transmit data and facsimile as well as private television programming. Customers use special decoders to view MDS transmissions, and the service operates as a common carrier similar to the telephone network.)

The MDS station was Micro TV, owned by Radio Broadcasting Company in Philadelphia. Micro TV borrowed from the British teletext system to create its own brand of teletext. Although a teletext service did not develop out of the tests, Micro TV did carry out experiments for some time and even investigated a form of two-way teletext. Subscribers could actually send text to the station to be broadcast by dialing the teletext computer, a Digital Equipment Corporation PDP 11/34. The computer could hold about two thousand pages (or screens) of information and accept up to eight simultaneous incoming feeds of digital information. Broadcast pages would reach adapted receivers at a rate of about six pages per second. Micro TV used lines 14–17 of the vertical blanking interval of its broadcasts of the Home Box Office (HBO) pay TV service, allegedly with only one complaint about interference with the HBO programs.

One of the results of the Micro TV experimentation was a proposal for a national teletext newspaper to be delivered via satellite. William Gross, the director of Micro TV, called the idea ''Info-Text'' and argued that a teletext newspaper could be delivered nationally to, say, four million viewers at a cost of three cents per receiver per month. [2]. The Info-Text concept led Micro TV to begin working with Southern Satellite Systems to use teletext to transmit data to cable television headends. (This system, which was eventually established without Micro TV, is described further in Chapter 4.)

Another early entrant in broadcast teletext, again with a special version, was KSL–TV in Salt Lake City, with what has come to be known as ''touchtone teletext.'' In 1977, KSL–TV, owned by Bonneville International Corporation, began engineering work in cooperation with Texas Instruments Corporation (which was already involved with teletext in England) to produce a teletext system in the United States. By the middle of 1978, KSL was on the air with teletext pages, using a system similar but not identical to the British teletext system. For one thing, the page format was slightly different, with 20 rows of 32 characters instead of 24 rows of 40 characters. The computer used for the system was a General Automation minicomputer, the same one already used for KSL's newsroom.

The most notable aspect of KSL's teletext trials was the development of the touchtone teletext system. In effect, a viewer could call the teletext computer and by pushing buttons on a touchtone keypad search the computer's data base for specific information. In practice, the system worked like this: the viewer watching normal teletext selects a specific teletext page by keying a number on the television's remote control pad; that page then displays a telephone number to call; the viewer picks up the phone and dials the number; the computer responds by placing an index (or menu) page into the teletext cycle using the same page number as the one the viewer is already looking at; the viewer sees the new index page and selects one of the topics in the index by pressing keys on the telephone keypad (the viewer is now interacting directly with the computer); the computer responds to each selection by finding the right page of information and slipping it into the broadcast cycle as a high-priority page. The key feature in teletext that makes this possible is that pages being viewed on a receiver can be automatically updated every time that page comes around in the broadcast cycle.

In this way, teletext is no longer merely a one-way system. The viewer can actually search through a massive store of information in a truly interactive mode. The viewer also experiences no real delay in receiving the pages of information because they are placed in the broadcast cycle as soon as they are found in the data base. The only limitations on the capability of the system are the number of input ports (i.e., the number of callers that can physically use the system at the same time) and the similar number of teletext "front" pages reserved for this activity.

Another aspect of KSL's teletext experimentation was the size of the magazine, or the total number of pages in the average broadcast cycle. The British systems tend to keep magazine size down to about 100 pages, with a total cycle time of about 25 seconds and an average or mean "waiting" period of 12 seconds (one-half the cycle time). The KSL system, however, at one period of the experimentation placed some 800 pages in the cycle, providing a cycle time of 120 seconds. Thus some viewers would have to wait about two minutes for a page to appear if the page selected had been broadcast just before the selection was made. As in the British system, though, the KSL system also tagged some pages with a higher priority so that these more popular pages would appear in the broadcast cycle more often. At other times during the trials, KSL kept the magazine size down to about 115 pages.

During the years that this was going on, KSL was studying the applications of teletext as well as the technical system. Among the many services suggested as possible teletext fare were the following [3]:

local supplement to television news
direct AP and UPI news wire
"perishable" advertising
schedules (movies, airlines, etc.)
weather watch
latest sports
periodic stock market listings
semiprivate pages (for special groups)
telephone directory
bank account information
transmission of computer programs.

Naturally, some of the suggested applications rely only on the one-way teletext technique, while other applications (using large computerized data bases) require the two-way link via telephone. KSL did not, however, test audience reception of teletext to any great degree, because during much of the time only two teletext-equipped television sets were available in all of Salt Lake City.

The KSL system was widely publicized in the United States by virtue of the distribution of a videotape depicting the system and by personal visits of industry representatives. The videotape was initially produced in 1978 for the National Science Foundation and the FCC, and over 2,000 copies were known to have been made. In addition, over 1,000 demonstrations were held at KSL's studios as of early 1980. Since then, Bonneville International Corporation has continued to examine teletext and in early 1982 began installing a small teletext system built by Zenith at KIRO–TV in Seattle, Washington.

Another early entrant in broadcast teletext was the CBS television network. CBS requested authorization from the FCC in early 1979 and began technical tests of teletext at KMOX–TV in St. Louis in March of that year. Unlike KSL and Micro TV, CBS was strictly testing the technical performance of the teletext signal. No data bases were created or used except for some test pages. Also unlike Micro TV and KSL, CBS did not simply adopt a known teletext system, such as the British system, but attempted to evaluate carefully the two then existing types of teletext: the British and the French systems.

Engineers from CBS spent time in both France and England to learn about the respective systems in their home countries, then acquired equipment from both places, modified it for U.S. television standards, and shipped the material to the CBS laboratories in Connecticut.

The modified teletext equipment, after additional work was done at the laboratory to make sure that all faults introduced by switching to the U.S. television standard were corrected, was then sent to St. Louis. There the origination equipment was installed at KMOX along with a microcomputer to control page transmission. A specially equipped truck traveled to various sites to make the field measurements of the teletext signals (see Table 3.1).

During the first part of the testing, there were obvious inequalities in the two systems. The first batch of British equipment could be set for only one data transmission rate, while the French equipment could be adjusted for five different data rates. Later, a new set of British equipment was obtained that could also provide five different data rates, although not quite the same rates. Also, the first set of British equipment produced pages with 20 rows of 32 characters, while the French system could display 20 rows of 40 characters. Again, the British equipment was subsequently changed also to display 40 characters on a line, but not until doubts had begun to set in regarding the flexibility of the British system. Another initial inequality, a factor of the test setup, was that the British system was allocated to lines 13 and 14, and the French system to lines 15 and 16, of the vertical blanking interval. Previously, lines 13 and 14 had not been used for broadcast teletext tests, and it was soon discovered (as suspected) that older television sets would experience interference with the normal programming if these lines were

Table 3.1. Measurements Made by CBS of the Teletext Signals in St. Louis

1. Signal strength
2. 2T Pulse response
3. Multiburst response
4. Signal-to-noise ratio
5. Subjective quality of picture
6. Interference analysis
7. Bit error rate
8. Eye pattern analysis
9. Signal strength margin of bit error rate
10. Teletext page evaluation
11. Signal strength margin of page errors

Source: Robert A. O'Connor, Teletext Field Tests, *IEEE Transactions on Consumer Electronics* CE–25(3): 307 (July 1979).

used. In a survey of KMOX personnel, 11 percent observed the presence of data on lines 13 and 14 on their home television sets (indicated by the presence of bright white dots at the top of the picture), and the trouble was traced to television sets using a 6GF7 tube [4]. For teletext reception, several types of modified TV sets were used, including a few Zenith and Sony sets. In general, the KMOX tests indicated that teletext was technically feasible using a European system modified for U.S. television, although a few problems remained. For example, in fringe reception areas where television pictures could still be seen, the teletext data may not be interpretable, and in areas where the ghosting effect afflicted the television picture, the teletext data may be corrupted. With regard to the line 13–14 problem, the tests' engineers simply stopped using those lines and put both systems on lines 15 and 16.

By the first part of 1980, CBS had reviewed the test results and concluded that the modified French system was the more flexible and thus more suitable one for U.S. teletext [5]. Moreover, CBS formally petitioned the FCC in July 1980 to adopt the CBS-modified Antiope system as the only standard for teletext in the United States. Following that move, CBS announced that it would establish a trial teletext service using the Antiope system on KNXT–TV in Los Angeles (this is described later in this chapter). In 1981, CBS executives let it be known that the CBS network, which had already carried a few teletext signals on a test basis, could be carrying a network teletext service by late 1982.

Finally, among the list of broadcast teletext pioneers is the Public Broadcasting Service and the line 21 system. As mentioned previously, the line 21 system is only a limited version of teletext in that the viewer cannot select pages. Nevertheless, the line 21 system developed a service and won a national audience before any other system (i.e., the closed-captioning service now reaching over 50,000 viewers with either special television sets or set-top decoders for normal television sets). In addition to closed captioning, the line 21 system can also provide text pages on two text "channels." Public television stations in Wisconsin, Minnesota, Nebraska, and California, along with PBS itself and the ABC television network, have begun using line 21 technology to broadcast pages of scrolling text for specific purposes.

Scrolling Text

In Wisconsin, WHA–TV has been a leading proponent of line 21 scrolling text and helped develop the software (the computer programs) to produce the scrolling text. The digital information is multiplexed into line 21 along with the digital signals for captioning and appears as text on a television screen when one of the text "channels" is selected. As a new line of text is broadcast and received, previous lines scroll toward the top of the screen. A full page of text occupies 15 rows of 32 characters. On television sets with built-in decoders, the text can appear in different colors, but on television sets using set-top decoders, the text display is only in black and white. Both the set-top decoders and the television sets with internal decoders are sold by Sears. (Sears is no longer the sole distributor; other stores will be selling these sets, which are assembled by Sanyo using Texas Instruments chips, or semiconductor devices, to detect the data and generate the appropriate characters.)

The two scrolling text services planned by WHA are news and health information for the hearing-impaired and agricultural information for the farming community. Eventually, WHA hopes to be the originating station for a network of public television stations in the state producing either a scrolling text service or perhaps a full teletext service using a different technology. The initial service, called Infotext, went into operation in June 1982.

In Minnesota, KWCM—TV in Appleton uses the two scrolling text ''channels'' for farm news and information, supplied by the Agricultural Marketing Service of the U.S. Department of Agriculture, the University of Minnesota, and the Chicago Futures Board. At least 250 farmers were known to be using the scrolling text service during the first months of operation in late 1981.

In Los Angeles, KCET–TV has been broadcasting a similar scrolling service but using the AP news wire and KCET's program listings as the content. This service, called ''Newsline,'' is presumably viewed by at least some of the estimated three thousand homes in southern California with line 21 decoders. In addition to Newsline, KCET broadcasts a full teletext service that is described later in this chapter.

In Nebraska, nine educational television stations led by KUON–TV have initiated a program to use line 21 scrolling text, primarily for agricultural information. The first information provider on the service was Agnet, an agricultural information network based at the University of Nebraska. The Agnet data are normally accessed via telephone lines, but the scrolling text service would extend usage free of charge to anyone with a line 21 decoder who is within range of the broadcast signals of one of the stations. (In the telephone-based service, Agnet has some 2,400 subscribers scattered across the United States and in some foreign countries.)

In a separate agriculture-related trial service, five PBS stations (in California, Colorado, Florida, Missouri, and North Dakota) are to begin in late 1982 a test of the Farm Market Infodata Service. This service will use the line 21 scrolling text procedure to broadcast agricultural marketing information provided by Agricultural Market News, a service of the U.S. Department of Agriculture that is normally transmitted via telephone lines to USDA offices around the country.

And in a nonagricultural application, WNET–TV in New York City began using the line 21 scrolling text procedure in 1982 to broadcast the Reuters news service within the WNET vertical blanking interval.

Two national networks, PBS and ABC, also use the text ''channels'' along with the closed captioning on line 21. The National Captioning Institute prepares a daily listing of captioned programs, along with plot summaries, that can be displayed as scrolling text. The service is called PLUS (program listing update service) and may eventually contain news, weather reports, and other information, scrolling by at a rate of four pages per minute. In addition, the National Captioning Institute prepares news summaries for the ABC evening news that appear on the second of the two text ''channels'' overlaying the video portion of the program when viewed. The NBC network carries line 21 closed captioning but does not carry the PLUS material. The CBS network does not carry any of the line 21 digital signals, arguing for the use of a full teletext system to handle subtitles and captions.

STV Teletext

One of the first stations in the United States to broadcast teletext was a special service station, as mentioned above. The application of the teletext technique to a subscription service, and specifically to subscription television (STV)—where viewers pay the broadcaster in order to receive an over-the-air television signal—has continued to attract attention as a viable vehicle for beginning teletext. In general, the perceived problem with teletext for the masses is that advertising revenue will have to cover the cost, but advertisers will not put money into teletext until masses of people already have teletext decoders. (Ultimately, this may be a false argument. In the case of color television, advertisers did pay for color productions before color sets were widely in use. This issue is discussed again in Chapter 7.) The way for teletext to begin, then, rests on procedures for covering the cost without relying on advertising revenue. A likely candidate is some form of subscription television where subscribers are already in the habit of paying a fee for the service and for the use of a set-top converter or descrambler.

Aside from Micro TV, another prominent company experimenting with subscription teletext has been Oak Communications, the largest STV broadcaster. In late 1980, Oak began preparations for a teletext test using WKID–TV in Ft. Lauderdale, Florida. The station normally broadcasts premium television programming (pay TV) under the name ON–TV, and the teletext signals were multiplexed into this signal.

Unlike all other teletext systems mentioned thus far, however, Oak did not place the data within the vertical blanking interval of the video signal but on an aural subcarrier. This had the advantage of enabling Oak to mix the technology for teletext decoding with that already used for descrambling the video programming. Although incidental to the teletext service, this use of the audio subcarrier also meant that data were transmitted in a relatively slow but steady stream, in contrast to the transmission of data in the vertical blanking interval, which is sent at high speeds but in discontinuous bursts.

Oak began broadcasting teletext pages in late 1980 and continued the trial until March 1981. Much of the system was designed and built by Oak, based loosely on the British teletext system. The screen format was 16 rows of 32 characters, because character-generating chips for that format were readily available; the computer used was an IBM Series 1 because another IBM Series 1 was already in use to control the pay TV programming. The teletext service carried information input at a terminal near the computer, relying on both local and national sources. Examples of national sources were UPI and the National Oceanic and Atmospheric Administration. The types of information displayed included news, weather, sports, farm information, consumer price information, recipes, health and first aid information, movie reviews, science features, and the like. Approximately 125 subscribers were provided with Oak descramblers that had been modified to decode the teletext data.

At the end of the trial, the users were surveyed and their reactions were generally not positive. Many found fault with technical problems in the equipment—an aspect Oak was well aware of. Other negative comments centered on the limited nature of a page (related to the admittedly restrictive format of only 16 lines of 32 characters), and the sometimes long wait for a page to appear. Some pages, for example, took as long as 30 seconds to appear, because that was the total cycle time for the approximately 60 pages in the magazine. (As with other systems, the number of pages in a cycle remained fairly constant throughout the broadcast day, but the content of the pages was changed periodically as groups of pages were shifted in and out of the broadcast cycle.) Despite the problems, Oak continues to believe that subscription teletext services have a future. This future, though, is not necessarily tied to the details of the Ft. Lauderdale service; Oak may decide to use a different screen format, the use of the aural subcarrier may be dropped in favor of using the vertical blanking interval, and the information itself may not be the mass market mix of the trial.

In addition to the WKID testing, Oak received FCC approval to broadcast teletext on STV station KBSC–TV in Glendale, California. Oak also has experimented with the teletext technique as part of cable television systems.

Another broadcasting company that supports the idea of subscription teletext, Cox

Broadcasting, purchased KDNL–TV in St. Louis with the intention of converting it to subscription television service. As part of that service, Cox has been investigating several types of teletext systems, including Oak's. Also like Oak, Cox has similar interests in text services on cable television and developed its own videotex system, Indax, for use on coaxial cable systems. Prior to developing an actual teletext service, whether on an STV station or a regular commercial television station, Cox reportedly surveyed television viewers and found that one-third would like to see some form of teletext service to access news, weather, and consumer information [6].

Given the subscriber-supported nature of STV and MDS broadcasting, it is not difficult to imagine the teletext technique being employed for a wide range of data transmission purposes as long as the over-the-air method is less expensive or less troublesome than wire- or cable-based delivery systems. At least one company, Star, Inc., of Santa Monica, California, has used a form of teletext on an MDS system to distribute text to hotels and motels. The trial service, established in Richmond, Virginia, used one line, line 25, of an MDS channel carrying pay TV and news on the video portion. The digital information multiplexed into the signal was composed of airline and other travel information. A single decoder was placed at a motel, where the digital information was decoded and placed on another channel as pages of rolling text.

Another MDS company, Microband Corporation, now owned by Tymnet, Inc., has also been playing with the idea of MDS teletext for several years. In 1979, Microband did some preliminary work with the French Antiope system and has continued to examine teletext as a way to distribute any kind of data on the local level (i.e., within range of an MDS transmitter and linked to national networks such as Tymnet's packet-switched data network).

Mass Market Trials

Beginning in 1981, broadcast teletext trials designed to assess the mass market potential for teletext services cropped up across the country. Almost all of them began broadcasting roughly the same mix of general interest information, but each also incorporated unique features. As a group, the various trials used just about every type of teletext system then available—French, British, Canadian, and U.S. These trial services began in Los Angeles, Chicago, and Washington, D.C., and later in San Francisco, Cincinnati, and Seattle.

Los Angeles

Los Angeles is the primary teletext location because of the number of television stations involved. The PBS station (KCET–TV), the CBS station (KNXT–TV), and the NBC station (KNBC–TV) are all taking part in broadcasting teletext services to a very small population of teletext-equipped television sets. The teletext efforts are also aided by the captioning services of the WGBH Caption Center, using the teletext system, not the line 21 method, to provide closed captions. The particular technology (and equipment) used by all these stations is the French, spurred largely by the fact that Télédiffusion de France lent more than $1 million worth of equipment to the trials.

The PBS station, KCET, had been interested in teletext for several years and had permitted teletext signals to be broadcast in a test mode as early as 1979. In 1980, the station received a grant of $100,000 from the Arthur Vining Davis Foundation to continue developing a teletext project, and in April 1981 KCET began broadcasting a teletext magazine called "Now." The stated purpose of the trial is to assess uses for teletext, and in particular, uses that stand apart from entertainment and advertiser-supported services [7].

As shown in Figure 3.1, the range of information on Now is indeed wide, from news and weather to a trivia quiz. But there are several features of an instructional or educational nature. One such feature supplements educational programs broadcast principally for in-school use. During the trial two programs, "Thinkabout" and "Inside/Out," were chosen for augmentation with textual material supplied via teletext. Students use a teletext-equipped television set in the school library or similar public place to step through additional pages of information. In the same vein, other KCET educational programs of more general interest are also supplemented with teletext pages, viewable by people with teletext decoders. (Approximately one hundred such sets are to be available in Los Angeles during the trial period; these sets are able to receive the teletext signals of any of the three stations broadcasting teletext services, because all use the same technology.) Quizzes are also part of the teletext service utilizing a feature of many teletext systems called "reveal." The "reveal" feature allows text to be present on the screen but blanked out until the viewer presses a "reveal" button on the decoder control. Other instructional pages are designed for children (the "Popsicle" pages), for

Figure 3.1. Sample pages from KCET's teletext service. *(Courtesy of KCET/Los Angeles Teletext Project, photo by Mitzi Trumbo.)*

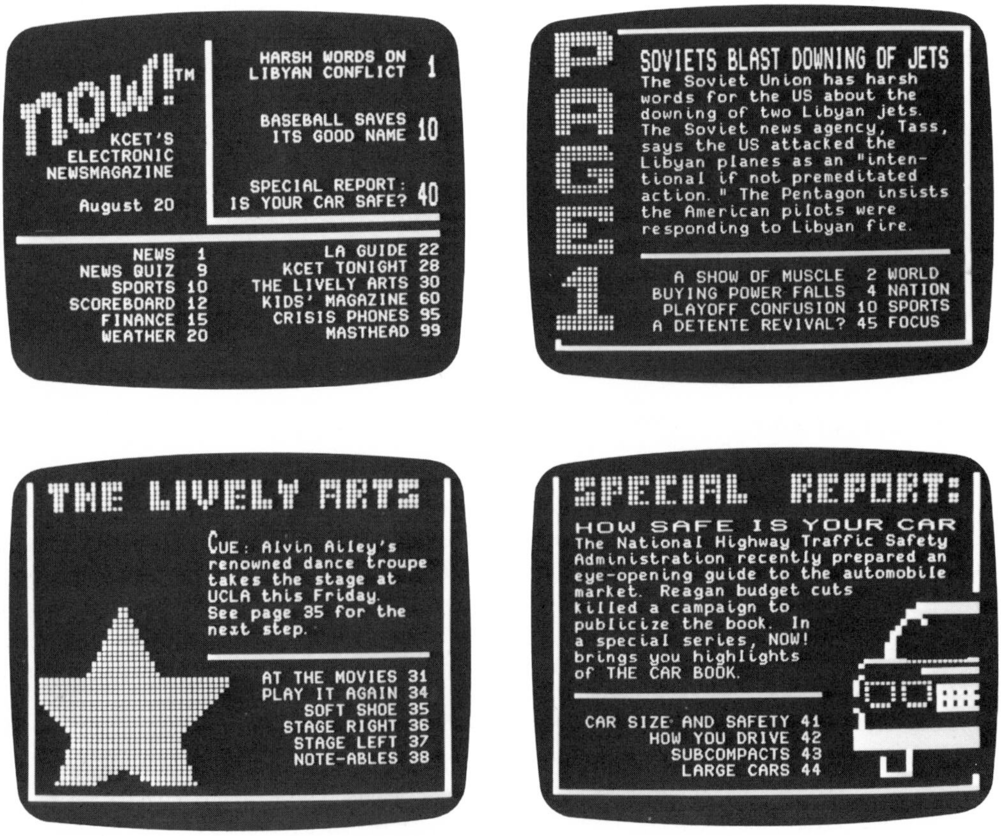

teachers (to read about upcoming instructional programming), and for schools (to receive administrative information).

Understandably, some of the experiences with the educational portions of Now have been less than satisfactory. With only a few sets in a school, teachers tended to use teletext as a group activity by leading a discussion that would rely upon or reinforce pages of text and graphics. But in these situations, students did not have individual control over the selection of pages, and a similar process could have been much more easily followed using a set of overhead displays instead of the teletext pages.

However, the ultimate educational benefits of teletext will be weighed when the trial service is completed in 1982 or 1983. During the trial, the educational applications are being guided by a steering committee of representatives from the Los Angeles Regional Educational Television Advisory Council, the Los Angeles Unified School District, and KCET.

KCET shares the teletext editorial center, the location where a Honeywell minicomputer is located as the hub of the teletext origination effort, with KNXT, the Los Angeles CBS station.

The CBS network had, of course, been experimenting with teletext for a few years before announcing in late 1980 that a teletext trial would be initiated at KNXT using the French Antiope system supported by CBS in the filing before the FCC.

Figure 3.2. Sample pages from KNXT's teletext service. *(Courtesy of CBS, Inc.)*

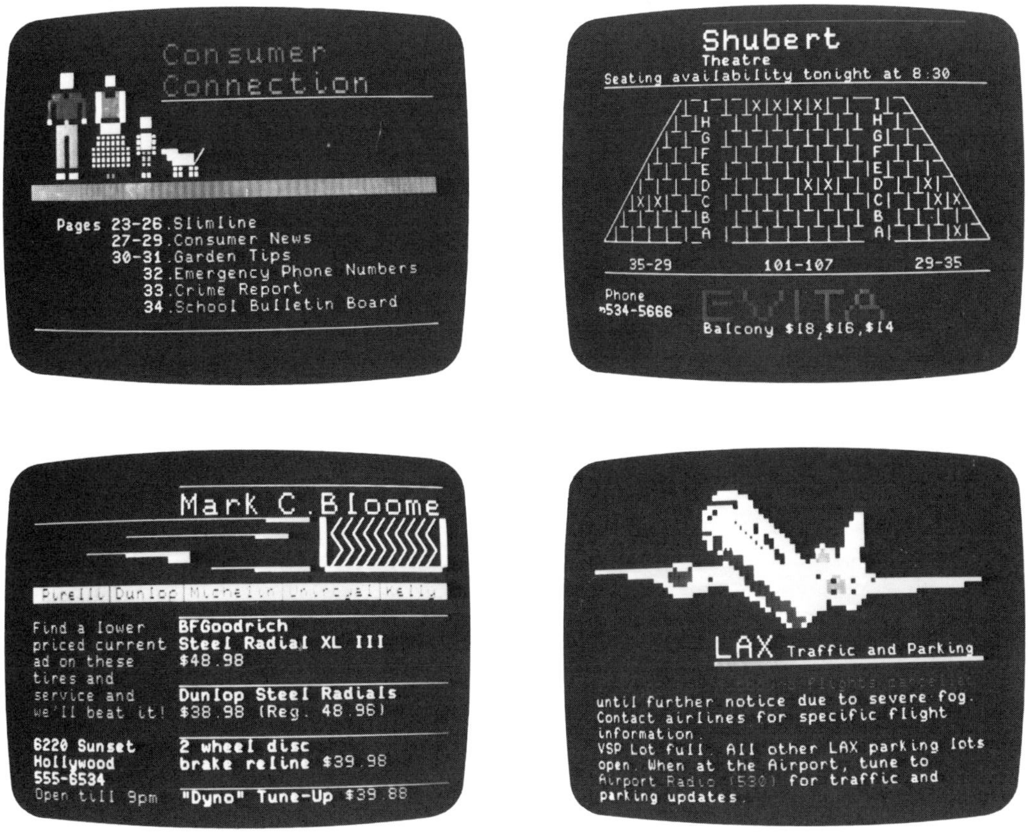

The KNXT magazine is called "Extravision" and, as does Now, contains a total of about 75 to 80 pages at any one time (see Figure 3.2). As Table 3.2 indicates, the broad range of general interest information covers news, sports, weather (including smog and ski reports), financial reports, travel information, and simulated advertising. The advertising is not "real" in the sense that, under KNXT's experimental teletext authorization, charging for advertising is not permitted. However, companies that indicated an interest in advertising on teletext and are taking part in the trial service include American Airlines, Coldwell Banker, Wilson Sporting Goods, Litton Industries, Merrill Lynch, and Ticketron.

Both KCET and KNXT (including the CBS organization) have worked closely together in what is essentially a joint project, even though both create their own separate teletext magazines. The stations share the same minicomputer and the same set of editing terminals and broadcast to the same set of teletext-equipped television sets, which are RCA sets modified by project engineers to incorporate teletext decoders. The screen format is 20 rows of 40 characters, and graphics are the "mosaic-six" style (i.e., any character-size space can be divided into six smaller squares, or two columns of three).

One of the additional features of KNXT's broadcasts is closed captioning, subtitled by the West Coast office of the WGBH Caption Center. As mentioned earlier, CBS has consistently refused to use line 21 captioning in the belief that a true teletext system offered better captioning capabilities. Thus the KNXT trial gave CBS a chance to demonstrate its position by hiring the WGBH Caption Center, which had been captioning programs for about ten years and had also experimented with the Antiope teletext system to prepare captions for selected CBS programs.

Table 3.2. Sample Index for KNXT's Teletext Magazine

Page	Contents	Page	Contents
1	Title	33	School Bulletin Board
2	Master index	34	On the move
3–8	News update	35–36	LA freeways
9	Weather index	37–38	Airlines/flight information
10	National weather map	39–41	Airlines/travel packages
11	LA today and tomorrow	42	LAX Traffic and Parking
12	Smog watch	43	Road advisories
13	Marine weather/ski report	44	Sports line
14	Recreational weather (mountains and deserts)	45	Sports briefs
		46–47	Sports day
15	Financial spotlight	48–49	At the track
16	Business briefs	50–52	Sports scoreboard
17	Dow Jones averages	53–62	$ales and $pecials
18	NYSE 10 most active	63	Entertainment spotlight
19	Amex 10 most active	64–65	LA calendar
20	Market diary (volume, advances, declines)	66	Weekly Variety's top ten
21	Consumer Connection	67–68	Today on 2 (TV schedule)
22–23	Slimline (diet menu)	69–74	Box office/Ticketron
24–25	Slimline (diet recipe)	75	Now on sale/Ticketron
26–28	Consumer News	76–77	Filmex
29–30	Garden Tips	78–79	Dining discoveries
31	Emergency Phones	80	Closed caption program guide
32	Crime Report	81	Help!

In late 1981, KNXT and KCET were concluding the first phase of the two-part trial. Both stations had teletext magazines on the air, but only about 15 teletext-equipped television sets were available for reception of the signals. The second part of the trial will see the deployment of the remaining 80 or so sets and will end in 1982 or 1983 with a market survey to examine consumer demand, teletext usage, and magazine content. The CBS network held a meeting for affiliated stations in September 1981 to discuss the trial and to caution stations not to leap too quickly into teletext services. On the other hand, CBS suggested that teletext could well be advertiser-supported by attracting the type of advertising that usually goes to newspapers (i.e., the current revenue base for television stations would not suffer). Regarding a national teletext service, which CBS announced in mid-1982, CBS officials have described a National Teletext Center that will feed teletext pages to the 200 affiliated stations. The affiliates will be able to add their own pages of local news and advertising. In addition, the CBS plan suggests that advertising agencies and other information providers or suppliers will have their own page creation equipment and will transmit finished pages to the local broadcasting station for insertion into the broadcast teletext magazine.

Rounding out the teletext picture in Los Angeles is the NBC station, a late arrival to the joint project. In mid-1981, after KCET and KNXT were already on the air with Now and Extravision, NBC announced that KNBC–TV would join in. The KNBC magazine is called "Tempo NBC Los Angeles," or simply Tempo, and contains the general interest mix of news, weather, sports, travel information and financial reports, plus items such as health tips, fashion information, and do-it-yourself information (see Table 3.3 and Figure 3.3). The KNBC pages were first broadcast in early 1982. Both NBC and CBS are funding a statistics-gathering method that uses meters in the individual teletext–television sets to record usage. And like CBS, NBC has announced that the Los Angeles teletext experience will form the basis for a national teletext service on the network. Both NBC and CBS also support a subset of the North American Broadcast Teletext Standard for transmission and display parameters (see Chapter 5 for more dis-

Figure 3.3. Sample page from KNBC's teletext service. *(Courtesy of the National Broadcasting Company.)*

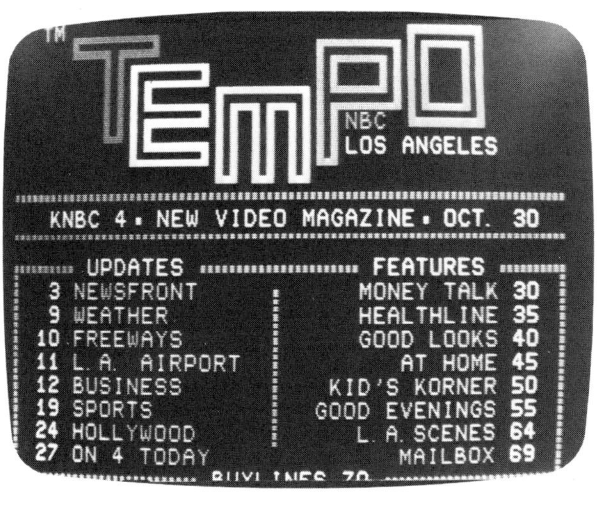

Table 3.3. Sample Contents for KNBC's Teletext Magazine

Page	Department	Page	Department
1	Inside Today	30–34	Money Talk personal finance tips
2	Contents page	35–39	Healthline personal health tips: exercise and diet information
3	Newsfront top news story	40–44	Good Looks fashion, beauty, and personal appearance information
4–8	Newsfront national, world, state and local news	45–49	At Home do-it-yourself information; tips on car care, plant care, and home decorating; recipes
9	Weather general forecast, air quality	50–54	Kids Korner word games, riddles, educational information on subjects of interest to children, such as astronomy
10	Freeways traffic conditions	55–63	Good Evenings movie reviews; theater, music, and restaurant listings
11	LA Airport traffic conditions, parking availability, map of airport	64–68	LA Scenes things to do in Los Angeles, including walking and biking tours, festivals, art galleries, and museums
12	Business top business story	69	Mailbox information on where to write "Tempo NBC Los Angeles"; viewers' letters published on Sunday
13–14	Business news briefs	70–78	Buylines full-page ads
15	Stock market summary		
16	Most active stocks NYSE		
17	Most active stocks Amex		
18	Market comment		
19	Sports		
20–21	Sports news briefs		
22–23	Sports scores and schedules		
24	Hollywood top entertainment story		
25	Hollywood news briefs		
26	Hollywood trivia quiz		
27	On 4 Today—listings		
28–29	KNBC program announcements		

cussion of the standards). The North American standard, based on Telidon and the AT&T videotex protocol, permits more finely defined graphics than is possible using mosaic-six graphics. In Los Angeles, NBC has demonstrated this higher level of graphics even though the Los Angeles teletext trials themselves use mosaic graphics.

Chicago

Chicago has been the second city of teletext. Although teletext in Chicago began being broadcast at almost exactly the same time (early 1981) as in Los Angeles, only two television stations are involved, and only one actually provides a teletext service. That one station is WFLD–TV, then owned by Field Enterprises. Probably it is a little unfair to Chicago to rate the city second to Los Angeles in teletext because the WFLD system in 1981 was larger and in some ways more sophisticated in several respects than the Los Angeles arrangement. But Los Angeles is a bigger television market, and three networks were involved, so Chicago has generally not attracted as much attention.

Field Enterprises became interested in teletext in early 1980 and decided to purchase the most complete teletext system then available. Because the system that had the most experience and was the most developed was the one in use at the BBC, Field personnel visited England, examined the system, and bought a similar one for use at WFLD. During the course of 1981, WFLD, through the teletext subsidiary of Field Electronic

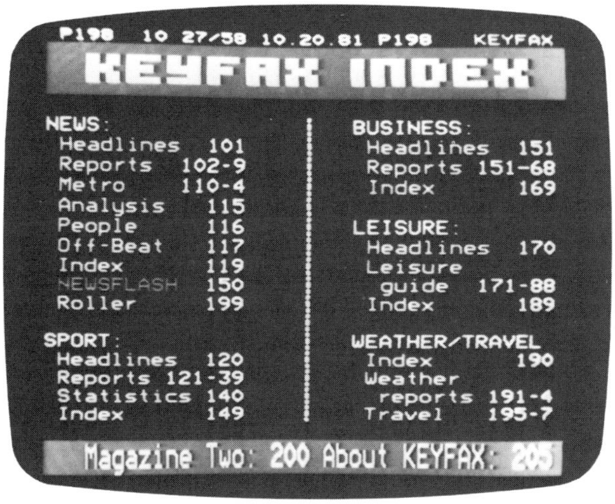

Figure 3.4. Sample page from WFLD's teletext service. *(Courtesy of Keycom Electronic Publishing.)*

Publishing, began broadcasting not only a teletext magazine but also teletext pages as normal video during otherwise off-air time periods (e.g., after midnight) and began testing teletext distribution to cable television headends for retransmission on the cable as normal video. (Field Electronic Publishing has since become Keycom Electronic Publishing, owned by the Centel Corporation. This change is addressed in a later section of this chapter.)

The teletext magazine, called "Keyfax," contains the usual mix of news, weather, sports, and entertainment information (see Figure 3.4). In addition, there are games and puzzles, horoscopes, and special interest items (see Table 3.4). The size of the magazine

Table 3.4. Sample Contents for WFLD's Teletext Magazine

Page	Content	Page	Content
100	General index		LEISURE
	NEWS	130	Guide
101	Headlines	131	TV programs
102–105	In detail	132	What's on
106	Analysis	133–135	Reviews
107	Metro	136	Top 10
108	People	137	Horoscopes
109	Off-Beat	138	Recipes
	SPORTS	139	Bull's eye
110	Headlines		WEATHER/TRAVEL
111–118	Results/reports	140	Guide
119	Calendar	141	U.S. weather map
	FINANCE	142	Chicago forecast
120	Headlines	143–145	Travel news
121–124	News/reports		OTHER PAGES
125	Dow Jones	146	News about Keyfax
126–127	Comex	147	Engineering test
128	Forex	148	Detailed index
129	Round-up	149	Rolling pages
		150	News flashes

varies throughout the day but is usually in the neighborhood of 80 to 100 pages. Some of the pages are known as rolling pages, which roll by without the viewer making a specific request. For example, if page 110 is a news story that is long enough to cover several screens, or pages, the pages of the story will automatically appear every 15 or 20 seconds once the viewer selects the first page in the rolling sequence. The amount of time that a rolling page stays on the television screen before the next page in the sequence appears is controlled by the editors at Field Electronic Publishing.

The second service that WFLD broadcasts, "Nite Owl," comes on the air between midnight and 6 A.M. (see Figure 3.5). This service is not really teletext because it is broadcast as a full video signal in place of any other video program. Viewers cannot select pages, but instead merely watch what amounts to a series of rolling pages. The pages are created on the teletext system and then converted to normal video before being broadcast. Everyone within reach of WFLD's signal can view the Nite Owl pages. In order to keep the sequence of rolling pages interesting and understandable, the pages are grouped into logical sequences called "orbits" that last 20 minutes each. In each orbit, some pages appear just to keep the viewer informed of what will be coming up in the next five or ten minutes, and other pages repeat information that is considered of high interest. For example, news, sports, and weather information appear briefly in every 20-minute segment, but each segment has its own predominant orientation, such as sports in one case, or leisure activities in another. The Nite Owl service is operated like a radio station in that an editor is on duty all night, not only to send out the latest news but also to broadcast messages to viewers, either generally to all viewers or specifically to an individual viewer after a request via telephone. Like radio, everyone tuned in is party to the comments of the editor/announcer.

Accompanying the Nite Owl pages of text and graphics is an audio background of "easy listening" music. This combination of text on the screen and unrelated audio was first authorized by the FCC for broadcast stations in October 1980; it applies only to the midnight-to-6 A.M. time period.

One of the unique aspects of WFLD's teletext and pseudoteletext services is that they

Figure 3.5. Sample page from Nite Owl, a nonteletext broadcast of teletextlike pages. *(Courtesy of Keycom Electronic Publishing.)*

are operated under a commercial authorization from the FCC. This means that WFLD can (and does in some cases) charge for the service through either the rental of teletext-equipped television sets or paid advertising. The teletext–television sets used in Chicago were manufactured specifically for that purpose by Zenith Radio Corporation; Field Electronic Publishing purchased about one hundred of the sets for distribution to viewers, both on a nonfee and a fee basis. During 1981, before all the sets were actually available, Field tried several rate structures ranging up to $4 per day. The second way to make money, paid advertising, is also in use, particularly on Nite Owl, because the latter service is viewable on all television sets. Advertising rates for Nite Owl ranged from $300 for a single night to $13,650 for a 13-week run. The type of advertising attracted by Nite Owl has been typical of late-night television—a large number of used car dealers.

Aside from the commercial side of WFLD's teletext, the technical arrangement also caused the Chicago venture to stand out. The system not only encompassed a teletext editing facility but also an automatic link to a major Chicago newspaper, the *Chicago Sun-Times,* also owned by Field Enterprises, tapping the newspaper's own computer system. In essence, Field Electronic Publishing established a teletext editorial setup using a Digital Equipment Corporation PDP 11/34 minicomputer system connected by telephone lines to the *Chicago Sun-Times's* computerized newspaper system called Atex, which also uses PDP 11/34 minicomputers. Physically, the teletext editorial facility is in suburban Chicago with two Atex computers and one teletext computer; the facility receives digital information from the newspaper's computers in downtown Chicago, then returns digital information to a teletext transmission computer at WFLD–TV, also in downtown Chicago (see Figure 3.6). A third teletext computer was for a time used as a backup transmission computer. Using the automatic link to Atex, the teletext editors can select stories to be transferred to the teletext system, edit them if necessary, and insert the pages into the broadcast cycle, whether as new pages or as replacements for existing pages.

The teletext system itself, like the one at the BBC, can hold thousands of pages in an online library, available to the editors at will, and can hold up to 16 separate magazines of a hundred pages each, which can be swapped in and out of the broadcast cycle either in response to immediate commands or under control of prestored commands. Also like the BBC's Ceefax, Field Electronic Publishing runs the service in a newsroom fashion, with a team of editors ready to put stories on the air as soon as they appear on wire services or via the Atex link. It is not surprising, then, that Field hired several ex-Ceefax editors to help run Keyfax and Nite Owl.

The appearance of the teletext pages in Chicago are nearly the same as in Los Angeles with regard to format and graphics. In Chicago, WFLD uses a screen format of 24 rows of 40 characters with the mosaic-six style of graphics.

In late 1981, Field Electronic Publishing began negotiations with several cable television systems to place teletext–television sets with cable subscribers. This would permit the Keyfax service to be seen outside WFLD's broadcast range and give the station an identifiable audience for the subsequent market research planned by Field.

In early 1982, Field Electronic Publishing became Keycom Electronic Publishing, a joint venture owned primarily by Centel Corporation, with minority ownership by Honeywell, Inc., and, to a lesser extent, Field Enterprises. The real effect of this change on WFLD's teletext service remains to be felt, as the old Field Electronic Publishing staff and system continue to function as before. WFLD–TV, which by itself is not part of Keycom, has continued to broadcast Keyfax and Nite Owl, although WFLD itself has

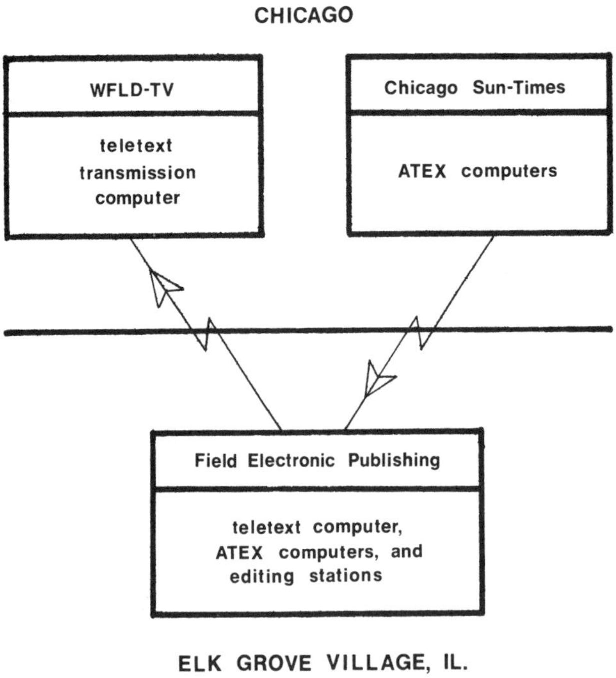

Figure 3.6. Diagram of the telecommunications links from the *Chicago Sun-Times* to the Field/Keycom teletext facility to WFLD–TV.

been sold to Metromedia and there have been rumors that WFLD would like to drop Nite Owl in favor of a normal video all-night news service, and consequently Nite Owl might move to the Chicago PBS station, WTTW. Aside from the teletext service, keycom's primary goal is to establish a videotex service using Honeywell terminals.

The second station in Chicago to broadcast teletext (on a far smaller scale) was WGN–TV, owned by the Tribune Company. In fact, WGN did not set out to produce a real teletext service but only a small series of pages for test purposes. Its microcomputer-based system was built by Zenith and modeled on the British technology. The notable aspect of WGN's entry into teletext was not the size, nor the announced plans to broadcast news, sports, and business data, but the fact that WGN is a superstation carried via satellite by United Video to cable television headends across the country. Specifically, WGN said that it intended to place teletext–television sets in Albuquerque, New Mexico, on a cable television system owned by the WGN group of companies. As previously described, however, United Video had its own plans for WGN's vertical blanking interval (i.e., as a means for disseminating the Dow Jones Newswire and other data to cable systems). In order to do so, United Video stripped off the WGN teletext data and inserted the Dow Jones data into the WGN signal.

WGN sued United Video for copyright infringement, claiming that the teletext data were related to the video programming, and specifically to the evening news. But in October 1981, WGN lost in district court, with the judge ruling that United Video was a passive carrier not ''performing'' the teletext data and thus not violating the Copyright Act of 1976. The court supported United Video's claim that if WGN wanted to use the

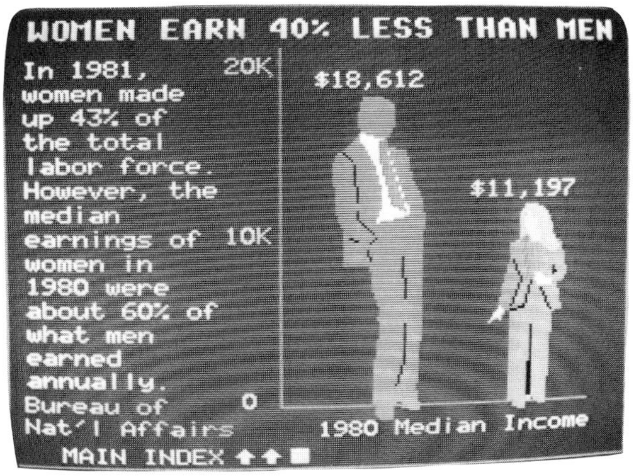

Figure 3.7. Sample page from WETA's teletext service. *(Courtesy of, and page designed by, the Alternate Media Center, Tisch School of the Arts, New York University.)*

vertical blanking interval to transmit data, the station should arrange for, and pay for, the transmission as a service wholly apart from carriage of WGN's video programming. The case left the ownership of the vertical blanking interval in a legal twilight zone, however because the case was actually resolved on copyright grounds. And to complicate matters, in late 1982 an Appeals Court reversed the earlier decision and said that United Video had to carry program-related teletext, but not necessarily other types of teletext. In turn, United Video has appealed that decision.

Washington, D.C.

In Washington, D.C., the teletext trial of WETA–TV is distinguishable from the other trials by the presence of public funding—the money to run the trial came directly from several federal agencies. Although the published objectives of the trial are to test the public's desire for teletext (if it exists) and not to test the technology, the WETA trial chose a system not in use anywhere else in the United States, namely, the Canadian Telidon system. (See Figure 3.7 for a sample WETA teletext page.)

The planning for the WETA trial got underway in earnest in 1980 primarily at the Alternate Media Center at New York University, under the sponsorship of the Corporation for Public Broadcasting and the National Telecommunications and Information Administration. At that time, the decision to use the Washington PBS station as the site for the trial had already been made, but no decision on the technical system to use was made public, although a published report seemed to favor the Canadian system over the French and especially over the British system [8].

In June 1980, the trial was officially announced as a project of the Alternate Media Center in association with WETA and funded by the Corporation for Public Broadcasting, the National Science Foundation, the National Telecommunications and Information Administration, and the U.S. Department of Education. Exactly how the publicly funded project went about selecting the Canadian system is not very clear, because there does not seem to have been an unbiased procedure for requesting and evaluating proposals. A news report at the time suggested that both the French and the British systems were never seriously considered, and that Zenith, along with the British, eventually pulled out of consideration because of objections to the way the selection process was being handled [9]. At any rate, backed by nearly $1 million in public funds and by the donation of several minicomputers from Digital Equipment Corporation, the Washington area trial was finally begun in mid-1981.

Information for the content of the teletext service was initially provided by two newspapers, the *Washington Post* (supplying news) and the New York *Daily News* (supplying sports information), and by a number of federal agencies: the National Weather Service, the Bureau of National Affairs, the Department of Education, the Federal Drug Administration, the General Services Administration, the Federal Trade Commission, and the Office of Education, among others. Pages are actually created in both the Washington area (the WETA offices are in Shirlington, Virginia) and in New York City. At the Alternate Media Center in New York, some of the pages of text and graphics are created on an editing terminal connected to its own PDP 11/03; the pages are subsequently transferred via dial-up telephone lines to WETA. For the pages created at WETA, the material for the pages comes from the information providers in a variety of ways, including by dial-up data links, facsimile, and mail. The general size of a magazine is about one hundred pages, similar to the magazines in trial elsewhere. In contrast to the other trials, WETA's Telidon system uses alphageometric graphics rather than the mosaic-six style.

Similar to the WFLD trial in Chicago, the WETA system is technically centered around a Digital Equipment Corporation PDP 11/34, although the machine at WETA is configured with less memory and less disk storage than the WFLD machine. A second minicomputer, a PDP 11/03, handles the actual arrangement and transmission of the magazine. The teletext-equipped receivers were purchased from Canadian sources and consist of Electrohome television sets with Norpak teletext decoders, and a cassette tape recorder for monitoring usage. About 50 to 60 receivers are to be placed in public and private locations. As of August 1981, the public teletext–television sets were available only in seven locations—a library, a college, two museums (including the Smithsonian), two community centers, and the National Press Club.

The objectives of the trial (see Table 3.5) are broad and comprehensive and may ultimately be met only when broadcast teletext services exist on a large scale. For example, the WETA trial hopes to be able to provide reliable data on the usage of teletext, on behavioral effects of that usage, on reactions to specific categories of information, and on economic factors affecting the service. Undoubtedly, given the small number of teletext–television sets in Washington and the fact that the service is a novelty for all viewers, the WETA trial will be able only to hint at possible answers, as will the other trials in Chicago and Los Angeles.

Table 3.5. Research Objectives for WETA's Teletext Trial

1. To provide reliable data on who uses the service and which pages are used most frequently.
2. To evaluate behavioral, attitudinal, and learning effects of teletext usage.
3. To decide if teletext can help people find information that they want and need.
4. To determine the popularity of the various segments of a teletext magazine.
5. To provide data on the costs of establishing and running a teletext service and the costs to the user.
6. To analyze management problems associated with establishing and operating a teletext service.

Based on: Teletext for Public Telecommunications Services: A Pilot Project on WETA–TV, Washington, D.C., *Viewtext 81,* pp. 84–85.

Although the Washington trial will not end until late 1982, a conference paper by one of the key participants in setting up the trial suggested some preliminary conclusions (of at least one researcher) [10]. The paper advised that broadcast teletext will need to complement or supplement the normal video programming so that the teletext "channel" does not compete with the host channel. The paper suggested further that for broadcast teletext to permeate the marketplace, a ready supply of set-top decoders must be available, and for technical reasons these decoders will perhaps be able to generate only alphanumeric characters in black and white. Although the paper did not say so, this has long been the approach adopted by PBS's line 21 system (i.e., the system carries no graphics, and set-top decoders display only black and white text). Beyond these concerns, the paper's author felt that broadcast teletext may be substantially, if not completely, eclipsed by text services on cable television, stemming from a host of reasons including lack of standardization for the broadcast signals, legal doubts over who owns the vertical blanking interval, and the current teletext characteristics of small magazines and sometimes slow responses.

Other Cities

By the middle of 1982, at least three other cities were preparing to begin broadcast teletext, again exhibiting the variety of technical systems chosen: in San Francisco, the French Antiope technology; in Cincinnati, a Logica (British) system; and in Seattle, a Zenith (U.S.) system.

In San Francisco, KPIX–TV, a Westinghouse station and a CBS affiliate, began preparations in late 1981 to expand the Los Angeles trials to northern California. The San Francisco trial is being conducted in cooperation with CBS and KCET in Los Angeles, and will use an Antiope system lent by Antiope Videotex Systems. Approximately 30 teletext-television sets will be placed around the city in homes and in public locations and will be moved periodically during the trial. By August 1982, the teletext service, called "Direct Vision," was formally on the air.

In Cincinnati, Ohio, Taft Broadcasting's WKRC–TV purchased a British-based teletext system from Logica, Inc., and Logica, Ltd. (the same group of companies that supplied the system to WFLD in Chicago). The system was a single PDP 11/34 minicomputer to handle origination and transmission of the teletext magazine. The teletext service, called Taftext, was scheduled to begin during the latter part of 1982, after technical testing that took place earlier in the year. Approximately 20 Zenith teletext–television sets were leased to Taft by the British under an arrangement that contains no charge for the first year of the lease.

In Seattle, Washington, an oldtimer in U.S. broadcast teletext—the Bonneville International Corporation, owners of KSL–TV in Salt Lake City—decided to begin teletext broadcasting on KIRO–TV, also under the same ownership. The system is a small Zenith one based on a microcomputer and nearly identical to the system used by WGN in Chicago.

In addition to these three announced trials, several other television stations had either begun some form of teletext or were preparing to do so in late 1982. In Knoxville, Tennessee, WBIR–TV began broadcasting teletext in conjunction with the World's Fair there during the summer of 1982. However, WBIR did not create the teletext pages. Instead, teletext pages created by Field Electronic Publishing (which during that time became Keycom) were transmitted by telephone line to Atlanta, where the digital information was multiplexed into the vertical blanking interval of superstation WTBS by Southern Satellite Systems. Because the WTBS signal is transmitted nationally to cable television systems, and the cable system in Knoxville was already carrying that signal, WBIR had only to tune to that channel on the cable system and, using a data bridge, pull the data out and place them into the WBIR vertical blanking interval. (Teletext in the cable environment is discussed at length in the next chapter.)

Other stations beginning to engage in teletext activity were WNBC–TV in New York City and WGBH–TV in Boston. While WGBH was preparing to use Antiope teletext equipment, the WNBC test broadcasts used equipment based on the newer North American Broadcast Teletext Standard.

Still other television stations across the country were also considering teletext in the early part of the 1980s but were hesitant because of the lack of standardization and doubts about any near-term return on the investment. At one time or another, over 20 stations either took part in limited technical testing or were prepared to do so, including WCBS–TV in New York; WRC–TV in Washington; KATV–TV, KOIN–TV, KGW–TV, and KTTV–TV, all in Portland, Oregon; KDNL–TV in St. Louis; WTVJ–TV in Miami, WITI–TV in Milwaukee; and WWLP–TV and WKEF–TV, both in Dayton, Ohio.

As this is being written, it seems that the future of broadcast teletext will be led in the early years by the work of the networks, particularly CBS and NBC. But the nature of broadcast teletext may also be determined by competition with similar services via other media, such as SCA, direct broadcast satellites, and cable television.

The teletext technique of multiplexing digital information into previously unused portions of a broadcast signal is not solely a child of traditional broadcast television. One of the other arenas for use of a similar technique is radio, thus giving rise to radio teletext. The more established name for this use of radio is SCA (Subsidiary Communications Authorization), or sometimes FM/SCA, because FM radio stations are the ones that engage in SCA transmissions [11]. Like the use of the vertical blanking interval, SCA uses spare capacity in the FM signal. This is a result of the fact that FM radio signals are granted a certain bandwidth by the FCC (i.e., each station is allocated 75 kHz of bandwidth) and the signal is modulated within the range. But because FM signals do not use the entire bandwidth, a subcarrier can be established and modulated to carry signals that do not interfere with the normal radio channel. But unlike the similar case in television, SCA has been used predominantly to transmit other audio signals rather than digital data. For example, one of the largest users of the SCA technique is Muzak and similar organizations, to deliver background music to stores, restaurants, and the like. Interestingly, the oldest use of SCA parallels the impetus behind teletext as a service to the handicapped. Whereas television's teletext grew out of efforts to aid the hearing-impaired, SCA has long been used in the form of radio reading services to assist the visually handicapped.

Two particular noteworthy applications of SCA in the 1970s were the Physicians' Radio Network, and a commodities market news service. The Physicians' Radio Network, started in 1975, was an attempt to provide medical information and news around the clock to doctors. An SCA receiver in a doctor's office was able to receive only that one service. The network was sponsored by pharmaceutical companies who provided receivers to doctors free of charge; drug advertisements, however, were part of the broadcast fare.

A second SCA service did in fact broadcast digital data to television sets. This service, by Market Information, Inc., of Omaha, Nebraska, used the SCA technique to broadcast commodities information to brokers and speculators beginning in 1976 [12]. The information comprising the content of the service came from the commodities exchanges and from Market Information itself. The subscriber needed a normal television set, a minicomputer, and the SCA receiver. Essentially, only four pages of text were broadcast, although the pages' content could change continuously and in this respect could be considered four magazines. Two of the pages displayed continuously changing price quotations (for grain and for livestock), and two pages displayed agricultural news.

Another SCA commodities service, called Market Monitor, is provided by Radio Data Systems, a division of Bonneville International. (Bonneville, of course, has also been heavily involved in teletext experimentation.)

A more recent attempt to use the SCA technique is Information Network Communications (INC), due to start in 1983. INC is a joint venture of National Public Radio (NPR) and a new company called National Information Utilities Corporation. INC will use the subcarrier of the NPR radio programs distributed nationally via satellite to carry digital information to be decoded and printed at special terminals. Although the actual INC services have not been identified yet, one suggested application is an information service for horse dealers. Other possible applications involve the one-way transmission of software to radio-receiver-equipped microcomputers.

Another relatively new SCA service is Dow Jones's Dow Alert, which uses analog subcarrier transmission techniques rather than digital. The service will provide business and industry reports as well as special features, and will be received by subscribers with special devices that can record specific programs for later playback.

In all, it is estimated that about one hundred FM radio stations across the country lease their subcarrier for SCA services of one form or another.

SCA has developed as a special service for specific audiences because that was the way the FCC authorized the service. An SCA receiver cannot tune to other signals; therefore the service operator can charge for the service by renting or leasing the receivers to subscribers, knowing that the receivers cannot be used for any other service. And although in principle SCA could be used for transmitting addressable messages to individual receivers, the FCC has to date maintained its position that SCA is intended to serve groups of listeners (or viewers, in the case of digitally fed television sets) with specific information needs.

LPTV is the acronym for low-power television, a limited television broadcasting station recently authorized by the FCC. Actually, low-power television has existed since the 1950s in the form of translator stations, which take a broadcast signal and rebroadcast it on a cleaner frequency. But in late 1978, the FCC began thinking of low-power television as a means of substantially increasing the number of television stations in the country, and in the process serving the various needs of small market segments. The LPTV stations, because of their low power, could exist in places where the number of traditional stations prohibited the addition of any new station with similar broadcasting power. The FCC subsequently announced that it would accept applications for approximately six hundred new LPTV stations.

During 1980 and early 1981, the FCC was flooded with some six thousand applications for low-power stations; many of the applicants intended to link the stations via satellite into networks, and some applicants specifically mentioned teletext as part of the proposed services. The types of programming listed in the applications varied from traditional television and subscription television to high-speed digital transmissions to business users. The Federal Express Corporation, for example, applied for a station to conduct subscription teletext in addition to video programming up until midnight of each day; after midnight the entire channel would be used for high-speed digital transmission of facsimile (i.e., transmission of "photocopies" of documents). Eventually, one can imagine the Federal Express Corporation engaging in national delivery of electronic mail, with documents transmitted by satellite to local low-power television stations and "broadcast" to addressable receivers.

Owing to the flood of applications, the FCC froze the process. So far, only one network of LPTV stations, educational stations in Alaska, have been authorized to begin broadcasting.

DBS refers to direct broadcast satellites, which can broadcast a television signal directly to small antennas at home or at work. DBS is another technology that has yet to get started in the United States, and it may change shape several times before DBS systems are put into place. The teletext potential of DBS is immediately present because

Table 3.6. Broadcast Teletext in the United States

Organization	City	Technical System	Approximate Year Started
PBS	(National)	Line 21	1973
Market Information	Omaha	SCA (radio teletext)	1976
Micro TV	Philadelphia	MDS, own (British-based)	1976
KSL–TV	Salt Lake City	Own (British-based)	1977
CBS, KMOX–TV	St. Louis	British, French tests	1979
PBS, ABC, NBC	(National)	Line 21 captions	1980
Oak, WKID–TV	Ft. Lauderdale	STV, own (British-based)	1980
KCET–TV	Los Angeles	French	1981
KNXT–TV	Los Angeles	French	1981
KNBC–TV	Los Angeles	French	1981
WFLD–TV	Chicago	British	1981
WGN–TV	Chicago	Zenith (British-based)	1981
WETA-TV	Washington, D.C.	Canadian	1981
KPIX–TV	San Francisco	French	1981
WHA–TV	Madison, Wis.	Line 21 scrolling	1981
KWCM–TV	Appleton, Minn.	Line 21 scrolling	1981
KCET–TV	Los Angeles	Line 21 scrolling	1981
KUON–TV	Lincoln, Neb.	Line 21 scrolling	1981
WKRC–TV	Cincinnati	British	1981
KIRO–TV	Seattle	Zenith (British-based)	1982
WBIR–TV	Knoxville	British	1982
WNBC–TV	New York	North American	1982
WGBH–TV	Boston	French	1982
CBS network	(National)	North American	1982-1983
NBC network	(National)	North American	1982-1983
WNET–TV	New York	Line 21 scrolling	1982
KMTF–TV	Fresno	Line 21 scrolling	1982-1983
KOZK–TV	Springfield, Mo.	Line 21 scrolling	1982-1983
WEDU–TV	Tampa	Line 21 scrolling	1982-1983
KFME–TV	Fargo	Line 21 scrolling	1982-1983
KRMA–TV	Denver	Line 21 scrolling	1982-1983
KRBK–TV	Sacramento	Not available	1982-1983

the television signal can carry multiplexed digital information. (Some time in the future the signal will probably be all digital anyway, but the "spare capacity" would still exist.) The Satellite Television Corporation, a subsidiary of Comsat, has proposed to carry teletext and closed captioning in its DBS signals, slated to begin in 1986 if events happen as scheduled. Exactly what the teletext service would be used for has not been stated, because the primary emphasis of this particular DBS system is on premium programming (i.e., pay TV).

Finally, one more facet of the broadcast teletext picture is VCR-teletext. This idea has been associated with teletext generally for a number of years, especially in France and to a lesser extent in England; it refers simply to the capability of teletext signals to control video recorders so that viewers do not have to be present during the actual teletext broadcast. This is something more than the similar idea of automatically videotaping normal programming, because the digital teletext signal may require a certain amount of time to build up a complex picture. Once recorded, though, the complex picture could be selected and viewed in full. Late at night, for example, the teletext broadcasts could be slowly building a store of information or complex pictures that would be held by the automatically controlled video cassette recorder.

In sum, broadcast teletext in the United States exhibits a varied and spotty picture. Here and there across the country, teletext trials have begun (summarized in Table 3.6), demonstrating that at least four or five different technical systems—with British, French, Canadian, or U.S. versions—can all do the same thing in pretty much the same way. No one is quite sure, however, how broadcast teletext will establish itself in the marketplace (or in the home, to be more accurate), with the possible exception of any service that is entertaining. Games and entertainment-teletext can probably attract enough viewers willing to pay, directly or indirectly, for the service. Secondarily, like the applications of SCA, broadcast teletext can find a certain amount of support in special services for limited audiences where subscription fees can provide the profit potential.

But given the uncertain nature of broadcast teletext's success (with possibly high receiver costs) and the growing percentage of cable television subscribers in this country, the text services on cable—and the applications of the teletext technique on cable—provide the second half of the description of television's teletext in the United States. The following chapter examines the development of teletext as a part of cable television systems for multiuse video, audio, and data communications.

References

1. Wells, Daniel R., PBS Captioning for the Deaf, Engineering Report no. E–7907, Public Broadcasting Service, 1979.
2. Gross, William S., Info-Text, Newspaper for the Future, *IEEE Transactions on Consumer Electronics* CE–25(3): 295–297 (July 1979).
3. Loveless, William, untitled presentation, in *Teletext and Viewdata Services* (Washington, D.C.: Society of Cable Television Engineers, 1980), pp. 15–24. See also Gary Robinson and William Loveless, " 'Touch-Tone' Teletext—A Combined Teletext-Viewdata System," *IEEE Transactions on Consumer Electronics* CE–25(3): 298–303 (July 1979).
4. O'Connor, Robert A., Teletext Field Tests, *IEEE Transactions on Consumer Electronics* CE–25(3): 309 (July 1979).
5. Statement by Dwight F. Morse, CBS Engineering, quoted in R. C. Morse, Videotext Explosion, *Cable Marketing,* May 1982, p. 84.

6. Cox Study Finds High Interest in Teletext, *International Videotex Teletext News,* January 1981, p. 9.
7. Goldman, Ronald J., Teletext for the Home and School at KCET, *E–ITV,* October 1981, pp. 39–41.
8. Teletext and Public Broadcasting, Alternate Media Center, New York University, April 1980.
9. But It Gets Underway Amidst Accusations, Charges of Unfair Policies, *International Videotex Teletext News,* July 1980, p. 7.
10. Gary Schober, The WETA Teletext Field Trial: Some Technical Concerns, paper presented at the Spring Conference of the IEEE on Consumer Electronics, June 3–4, 1981.
11. See, for example, Eric Somers, The Growing Potential of FM/SCA, *Radioactive,* February 1979, pp. 8–9.
12. Eric Somers, Digital Broadcasting "Piggy Backing" FM Subchannels, *Broadcasting Systems and Operations,* April 1979.

4

Cable's Teletext

Cable television systems were originally constructed as simply a means of getting a broadcast television signal to homes in poor-reception areas. From the late 1940s to the mid-1960s, cable television systems did just that, and the total number of subscribers (by 1964) was fewer than one million, scattered among some 1,200 systems across the country. But beginning in the 1960s, talk of computer utility companies and widespread data networks led cable television into the role of digital data distributor as well as television signal distributor, because, with a bit of work, the coaxial cable of the cable television system could compete with the wires of the telephone company. Moreover, the broadband coaxial cables had an immediate size advantage over the narrowband telephone wires, enabling the cable to carry more data faster.

As cable systems developed, a variety of two-way video, voice, and data communications were tested in the early and middle 1970s, but the economic climate was not favorable for the development of services using these technologies. Nevertheless, the basic concept of cable television as an information and communication system continued to be explored, and a variety of techniques have been proposed and implemented, including teletext on cable. Thus quite apart from broadcast teletext (which can certainly be carried by a cable system that receives the "host" station), the teletext technique has been adopted by cable as a means of delivering digital information to cable headends and to cable homes.

But in order to set cable teletext in context, it might be helpful first to describe other techniques used by cable television to deliver nontraditional services (e.g., "billboard channels'' and digital network procedures). In one way or another, all the nontraditional cable services that largely display text on a television screen have been called ''cabletext,'' but the techniques used to deliver the data to the screen are quite different from case to case and have played a major role in shaping the respective services.

Billboard Channels

Probably the simplest form of cabletext is the billboard channel, seen on many cable television systems. Generally speaking, these channels display alphanumeric characters on color backgrounds, with the text scrolling by in one or more directions. For example, the channel may display a screen split into sections for different categories of information, with text appearing as rolling pages in one section and as a left-to-right crawl in another section (see Figure 4.1). The text is generated by a character generator at some central location such as the individual cable system's headend, and the resulting video signal is just like any other television signal on the cable, occupying a 6-MHz band width.

Typically, the data that feed the billboard channel (or character-generator channel, as it is sometimes called) arrives at the cable headend as a digital data stream. Some of the major providers of this digital information are the Associated Press, United Press International, Reuters, and the National Oceanic and Atmospheric Administration (the latter for weather data). Reuters may have been the first news organization to create a news service specifically tailored for display as a text channel on cable television, beginning in 1970. The Reuters cable news service, called News-View, is delivered to

Figure 4.1. Sample billboard or ''cabletext'' page. *(Courtesy of Beston Electronics, Inc.)*

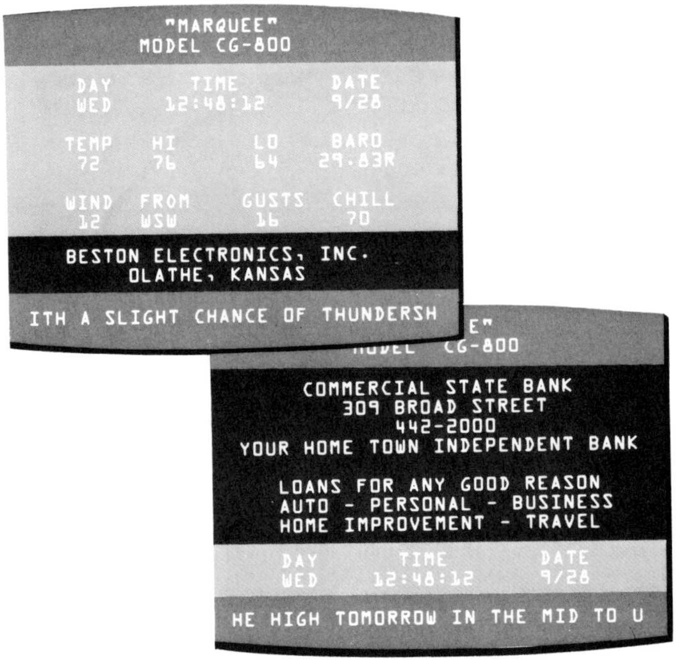

cable systems by both common carrier land lines and satellites. (The latter method is discussed later in this chapter as an application of the teletext technique.) The UPI service, called Cable Newswire, provides constantly updated news, weather, and sports arranged in a 15-minute cycle for display as rolling pages of text. The AP Newscable service provides a similar mix of information with a format for display on a television screen; additionally, AP provides a state news service, the Washington executive news report, a stock market and economic report, and Spanish language reports, although these do not have a format for direct display as cable television pages. The Associated Press has also developed an AP viewdata service based on videotex experiments in Coral Gables, Florida, and Dallas, Texas. The service is specifically designed for teletext distribution via broadcast or cable television, or for videotex distribution by telephone lines or narrowband cable channel. The average story length is 400 to 500 words divided into lines of not more than 40 characters each. Stories can be transmitted by the AP editor in New York City with one of eight levels of priority and coded for at least 40 categories of interest. Some of the categories are listed in Table 4.1. In a given day, the AP viewdata service may distribute stories with a combined total of some 80,000 words.

Aside from these traditional billboard channels, cable systems have more recently begun to be used for text channels programmed by a local newspaper or similar publisher. A newspaper may lease a channel or two from the cable system and, using a character generator and some associated equipment, compose pages of local news, sports, community events, and even classified ads. As billboard channels, of course, the video signal distributed from the cable headend is a normal television signal, and viewers must wait for lines or pages of text to scroll, roll, or crawl by.

Table 4.1. News Categories Used by the Associated Press for Videotex, Teletext or Cabletext Services

Category	Code
World news	1100
National news	1110
People	1120
Latin America	1130
Politics	1150
Commentary	1160
Weather	1300
NYSE stock table	1350
General business news	1390
General markets	1400
General sports	1500
Entertainment	1620
Style and trends	1630
Travel, health, science	1860
Home	1870
Baseball	2601
Basketball	2602
Hockey	2603
Football (pro)	2606
Soccer	2607
Newsbrief	1000
Nonprintable advisories	9999

Source: Associated Press, Trying to Cope with Videotex, Teletext and Viewdata?, April 1981.

One of the first newspapers actively to promote this form of electronic publishing was the *Daily Sun* in Yuma, Arizona. Beginning in mid-1980, the newspaper leased four channels on the local cable system and included the AP international news along with the newspaper's own items. The channel devoted to local news is updated four or five times a day and carries a total of about 60 pages at any one time—two-thirds news and one-third advertisements. Some of the rather brief news items contain references to the more complete treatment in the *Daily Sun*. The channel devoted to classified ads contains a like number of pages rotating in a similar 15- or 20-minute cycle. In 1981, the *Daily Sun* charged $15 per day for a display ad and $7 per day for a classified ad on the cable channels, and asserted that the cable news service was beginning to pay its way.

Other newspapers across the country have also bought character generators and leased cable channels, although the actual numbers involved are rather low. A March 1981 survey by the National Cable Television Association listed only ten newspaper/cable services, with the oldest being the *Murray Ledger Times* on the Murray Cablevision system in Murray and Mayfield, Kentucky, since 1979 [1]. In Murray, a single channel is used for displaying a one-hour cycle containing 25 percent local news and 75 percent AP national and international news mixed with local ads. By mid-1982, the number of newspaper/cable services was expected to begin growing, with an estimated 30 to 40 newspapers involved. At least one major newspaper chain has advised member papers to enter the newspaper-on-cable business using character-generator systems.

Aside from these billboard channels of text, there are many variations of the text channel that have appeared on cable systems. One idea that might be singled out is the mixture of automatically generated text with moving video, whereby a screen of text describes certain items or stories; then, perhaps in response to a digital signal from a subscriber, the text is replaced with a normal television segment. This concept has been behind cable television experiments for at least a decade, but it is just beginning to appear to be economically feasible. One such application is a trial service jointly sponsored by the J. Walter Thompson advertising agency and the Adams-Russell cable television company. The trial service is called Cableshop. It permits viewers to telephone the cable system operator to request specific short video segments describing products. On a billboard-type channel, the viewer sees a constantly updated schedule of the segments already ordered for the next half hour, to be shown on other channels (see Table 4.2).

Table 4.2. Sample Items on a Cableshop Menu Page

Code	Message	Time	Channel
103	Cook with Alum Wrap	now	21
203	Solar Heat Now	now	22
104	Dress for Success	9:35	21
204	House Buying Tips	9:35	22
105	Super Supper Quiche	9:40	21
205	Town Meeting—Budget	9:40	22
101	Better Putting	9:45	21
201	Adult Ed Courses	9:50	22
102	Build a Tool Shed	9:57	21
202	High Yield Bonds	9:59	22
103	Cook with Alum Wrap	10:07	21

The computer controlling the Cableshop channels tries to schedule the video segments so that a viewer will not wait more than 20 minutes; if that is not possible, the computer advises the viewer to call later. In future installations, a computer/controller may be required for every 3,000 to 4,000 cable subscribers. The J. Walter Thompson agency is evaluating the current system in part on the basis of knowing who is watching what, because each subscriber has a unique access code.

As another example, the Montreal cable system has a similar service to provide "documents" or short programs on request. One text channel contains a listing of upcoming documents or programs; viewers telephone their requests to a cable system operator who takes the requested video cassette and places it on a videotape player. A computer handles the scheduling and generates the schedule displayed on the text channel.

Various other schemes have been tried for giving viewers some measure of control over the specific video items received, and Warner Amex has suggested that its Qube system might evolve into a video-on-demand system.

All of the billboard-type channels, however, involve no special equipment between the cable origination point and the home television. The screen display is distributed as a standard television signal and is viewed at home on a full channel dedicated to that service.

Data on Cable

There are other ways of producing similar displays on home television screens that involve digital data transmission and equipment in the home to receive and perhaps process the data. Among these procedures is use of the teletext technique, but it might be best to describe first some of the other ways of carrying digital signals on the cable system, either to all viewers or selectively to individual viewers.

For a number of years, cable systems normally devoted to the distribution of television signals have been carrying digital signals for a range of data communications applications, although only as the 1980s began and the concept of local area networks gained favor did the cable television industry start becoming serious about digital communication.

One of the characteristics of the cable television system has been its lack of a switching mechanism. Unlike the telephone system with exchanges full of switches to connect any single caller with any other single telephone, cable television systems simply distributed the same set of channels from a central point to all receivers. If you wanted to send signals to each subscriber individually on, say, a 10,000-subscriber cable system, you would have to take some on the frequency space on the cable and divide it into 10,000 narrow channels (frequency division multiplexing), or you would have to take one channel or set of frequencies and "talk" to each subscriber in turn, one after the other (time division multiplexing), or a combination of the two. This is, in fact, what has been done to permit individual communications on cable systems, and one procedure using these techniques is called polling. In the polling process, a control computer interrogates subscriber terminals in turn, and at the time of each interrogation the home terminal is allowed to respond with a digital message, which might contain an identification tag and a small amount of data. The data can be the actual message to be transmitted, or a request by the terminal to move to another less used channel to transmit a greater volume of data. Beginning in 1971, polling techniques were utilized in a number of consumer-oriented two-way cable television experiments, including tests by Sterling Manhattan Cable Television in New York City, Telecable Corporation in Overland

Park, Kansas, ATC Corporation in Orlando, Florida, and United Cable Television Corporation in Carpentersville, Illinois [2].

The polling technique has also been at the heart of the Qube system developed by Warner Amex Cable Communications and used in Columbus, Ohio, since 1977. (Qube has recently begun to be installed on other cable systems in other cities.) Although the Qube system, as a complete cable services package, includes provisions for selecting pay TV channels and the like, the digital interaction accomplished using current polling procedures permits subscribers to send a single digit (from the numbers 1–5) to the cable system's computer. For example, if an announcer on the screen asks a question, the viewer can respond by pressing "1" for a yes, or "2" for a no, or make multiple-choice answers. This exemplifies a basic limitation of current polling techniques when used for cable systems serving tens of thousands of subscribers—it takes time to poll every terminal, and this limits the amount of data that can be returned to the computer in a given amount of time. Because of this, Warner Amex in Columbus chose to use the interactive digital portion of Qube primarily for asking simple questions of the audience in conjunction with talk shows, sports events, and similar suitable types of entertainment. The result has been understandably less than enthusiastic use by home viewers once the novelty effect wears off [3].

One of the applications that the polling technique does seem suited for is status or alarm monitoring. Sensory devices attached throughout a cable system are polled periodically by computer to ascertain a change in the conditions being monitored and the digital identification of the monitoring unit responding. For home use, the monitors include fire, intrusion, and medical alarms and a channel-being-viewed monitor to permit pay-per-program services. For business and municipal use, the monitors include fire and security alarms, utility meter reading, and traffic flow monitors. For the cable company itself, polled monitors can be used to ensure that the entire physical plant is operating correctly. A typically polling system employs a message structure that allows only eight or ten bits of data to be returned to the computer on each query (i.e., eight or ten sensory devices could be tagged as on or off). Such a system using one narrowband channel for polling and one channel for receiving messages could require up to six seconds to poll sequentially only 4,000 subscribers.

A second technique for handling digital communications on cable television systems is known as the contention method, or CSMA/CD (carrier sense multiple access, collision detection). This has often been explained as something like a conversation at a party—while someone talks the others listen, but as two or more people start to talk at the same time, the speakers stop and someone starts again, with one usually managing to start before the others. On the cable system, all the terminals use the same bandwidth to transmit information to a central computer, but a terminal does not transmit if another terminal is already doing so (transmissions are broken into packets so that no terminal monopolizes the channel). Once the channel is clear, a terminal may transmit; if two or more terminals start at the same time, each stops and restarts after a random delay, which usually assures that one restarts before the other(s).

The contention technique is itself part of a broader group of technologies developed for high-speed data communications among machines located geographically rather close to each other. These systems, called local area networks, are just beginning to be installed within the data processing industry. Not all local area networks use the CSMA/CD method, and it is likely that by 1983 or 1984 the Institute of Electrical and Electronics Engineers will suggest three standard forms of local area networks: CSMA/CD using baseband signals; a token-passing system in a ring structure using twisted wire pairs; and a second token-passing system designed for the coaxial cable of cable television systems. (A token-passing system involves the use of a unique code, or token, that travels the network and must be grabbed by a machine in order to transmit on the channel. After a message is sent and received, the receiving machine releases the token to be available again on the channel.)

As an example of this type of data communication in the cable television environment, Cox Cable Communications has developed their own system using the contention technique, called Indax (interactive data exchange). A portion of the cable's bandwidth is dedicated to interactive communications at 28 kilobits per second using CSMA/CD procedures. Interestingly, another part of Indax uses the teletext concept to deliver data to viewers by transmitting pages of text and graphics in digital form in cycles on certain subchannels. As with teletext, the viewer must wait for a requested page to appear in the cycle before the page will appear on the screen. However, the data are not multiplexed into a television signal but transmitted at 28 kilobits per second as a digital signal on their own subchannel.

A third technique for carrying data on cable can be much simpler than polling or local area network techniques, and that is to dedicate a portion of the bandwidth for digital communication between two points, and to handle the data communications just as if a telephone line were connecting the two points. Similar to a telephone-based system, the cable system in this instance uses a modem (modulator/demodulator) as the interface between the computer or terminal and the transmission line. In the case of cable television, the modem is an rf (radio frequency) modem, designed to work at radio frequencies. Beginning in the mid-1970s, a small number of cable television systems have leased channel space to businesses for point-to-point data communications on the dedicated frequency using rf modems. Manhattan Cable Television claims to be the first cable television system to carry digital traffic for business customers, beginning with interconnections among several banks in 1975.

Using these techniques either alone or in combination, cable television systems have continued to examine the delivery of data services to all the homes on a cable system and, in effect, to replace the telephone as the means of connecting a home computer terminal or microcomputer with a central computer. Dow Jones and Company, which offers several computerized data bases for business and home use, has for example been working since the latter part of the 1970s with several cable systems to jointly develop the necessary hardware for a cable subscriber with a microcomputer to "call" the Dow Jones News Retrieval Service via the cable system rather than the telephone network. In addition to the equipment needed on the local level, the digital communication has to be relayed to Dow Jones's central computer complex. This can be accomplished by Dow Jones's own satellite network, currently used to deliver the text of the *Wall Street Journal* to distant printing plants. Similarly, Warner Amex has begun enhancing its Qube system for home microcomputers to "call" distant data bases.

These forms of data on cable are just emerging in the marketplace. However, for cable systems operators in the early 1980s the more important question is not data communications but rather addressability—how best to install systems that permit a central computer to contact each subscriber for such mundane applications as turning service on and off, authorizing pay TV channels, changing the authorizations for dozens of levels (or tiers) of premium programming, and detecting a malfunctioning or disconnected home terminal. The addressability problem can be solved by a variety of techniques, from polling to teletext, and already a half dozen or so cable equipment manufacturers offer addressable converters (e.g., Oak Communications, Tocom, Inc., Jerrold

Division of General Instruments, Scientific-Atlanta, Inc., Zenith Radio Corporation, Pioneer Communications). The converter is the channel selector box that sits on top of the subscriber's television set in many cable systems. The issue for the cable industry is not so much how addressability is accomplished but what the resulting system can do—and in the early 1980s what cable operators want to be able to do is to communicate economically with the converter in each home for simple maintenance tasks. The introduction of addressable converters on a large scale is likely to precede the widespread use of cable television systems for data communications, with the exception of the relatively small number of cable subscribers with microcomputers or similar terminals who, on certain cable systems, will be able to use the cable system in lieu of the telephone network.

Satellite Delivery

Within the range of digital communications techniques in the cable television environment, there is a place for teletext as long as it serves a need (i.e., permitting a cable operator to increase or enhance service), whether from the operator's or the subscriber's viewpoint. One of the applications that has taken hold is the use of the teletext technique to distribute text and graphics nationally via satellite to cable television headends. This application follows from the fact that in recent years television signals have been nationally delivered by satellite to cable systems to bring premium programming or pay TV, the superstations, and similar nonbroadcast fare to the cable subscriber. Because the teletext technique involves multiplexing data into an existing television signal, the television signals already being distributed could obviously carry some additional information. In this manner, a digital signal can be distributed to cable system headends without the trouble of arranging a separate channel on the satellite or arranging for the use of land lines. The first television signals to be thus exploited were the superstations, such as WGN in Chicago and WTBS in Atlanta, whose signals are carried by satellite to thousands of cable television systems across the country.

One of the first companies to develop teletext for satellite delivery of data to cable television systems was Southern Satellite Systems of Tulsa, Oklahoma. Southern Satellite is responsible for distributing WTBS–TV (and a few other television signals) via RCA's Satcom I satellite, and the signal is received by over 3,000 cable systems. In 1979, Southern Satellite began working with Micro TV, a teletext pioneer in the United States, to utilize the WTBS vertical blanking interval for teletext purposes. After a short time, Southern Satellite turned to Zenith Radio Corporation for additional help, and Zenith eventually produced the system that Southern Satellite put into place in 1980.

The service provided by Southern Satellite is called CableText, an umbrella term for the various types of transmitted data. Initially, the cable news services of UPI and Reuters were the only transmissions, but in late 1981 Southern Satellite began adding a stock and commodities news service (Vector Consumer News) and a weather service (View Weather). About 80 cable headends had installed the necessary equipment to pull the data out of the vertical blanking interval and generate the pages of text and graphics.

The decoding of the teletext signal at the cable headend or studio was initially accomplished using a Zenith decoder/character generator, although the ability to connect to other manufacturers' equipment was later added. The Zenith system produces a display of 19 rows of up to 40 characters each, using the mosaic-six graphics format, based upon the British teletext systems. Because the Zenith unit generates a visual display to be shown on an otherwise unused channel, the end result for the home viewer is another billboard channel of rolling pages with some graphics capability. For the cable operator, however, the advantages include elimination of land lines to receive the digital information and the fact that the headend equipment is itself addressable (i.e., tailored or individual information is picked up only by the designated cable system or systems). For example, the View Weather service can provide region-specific forecasts even though the signal is nationally distributed by the satellite, because portions of the service can be addressed to the proper parts of the country.

In addition to the Zenith unit, the CableText service can be received by other types of character generators and even microcomputers and similar computer equipment, as long as the appropriate decoder is used. By extension, if the data are stored at a cable headend or control point, an interactive information service could be created between the subscriber's terminal and the cable system's computer, with updates to the stored information being received via CableText.

In mid-1982, several developments brought broadcast teletext and cable teletext closer together. As mentioned previously, at about the same time that Field Electronic Publishing became Keycom Electronic Publishing, it was involved in several efforts to supply teletext pages to cable television systems. Under one plan, Keycom will deliver teletext page data to Southern Satellite Systems (owned by Satellite Syndicated Systems) for national distribution as part of Southern Satellite's existing CableText service (see Figure 4.2). In the same vein, Keycom pages were delivered to WBIR–TV in Knoxville, Tennessee, by virtue of being sent by telephone line to Southern Satellite Systems outside of Atlanta to be inserted into the WTBS signal, which is picked up by the Knoxville cable system. During the Videotex 82 conference in New York City in June 1982, Keycom teletext pages were similarly retrieved from the WTBS channel on the Manhattan cable television system serving the conference site. Keycom also intends to begin a telephone-based videotex service called Keytran that will use the North American standard for page displays, not the display and transmission standards currently used for Keycom's teletext.

Another company in much the same situation as Southern Satellite is United Video, Inc., which transmits another superstation, WGN–TV, via satellite to cable systems. As mentioned previously, United Video and WGN each had their own plans for that vertical

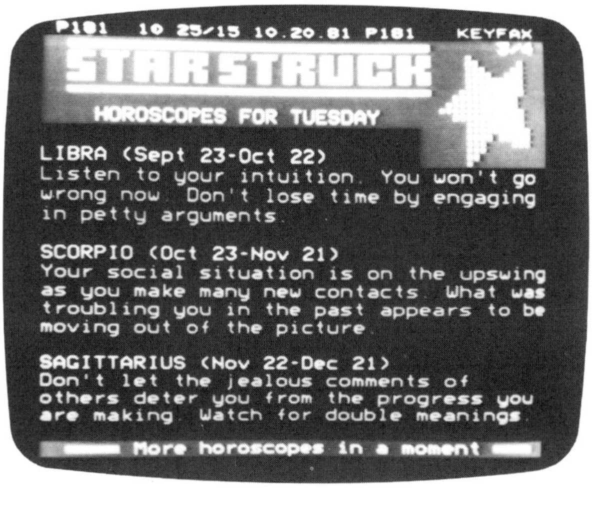

Figure 4.2. Sample page from Southern Satellite Systems' CableText service. The Keyfax page shown is from the joint venture between Keycom Electronic Publishing and Satellite Syndicated Systems. (*Courtesy of Satellite Syndicated Systems.*)

blanking interval, and after going to court, United Video won and then partially lost the right to decide what data, if any, will occupy the WGN vertical blanking interval once the signal passes into United Video's hands.

Although United Video's use of the WGN signal is much the same as Southern Satellite's use of the WTBS signal, a few differences can be mentioned. United Video does not use the Zenith-developed British-based teletext system, but rather the French-developed teletext system. Using the French Didon protocol for teletext data transmission, United Video transmits the Dow Jones cable news service to appropriate decoders at cable headends. On the service side, United Video has introduced a continuously updated program guide, called the Electronic Program Guide, that can be individually prepared for subscribing cable systems. At the cable site, the digital information is decoded and a display is generated that constantly provides the schedule for the next two-and-one-half hours of programming on the cable throughout the day. Again, from the viewer's viewpoint the schedule appears as another billboard channel and is not true teletext. Separately, WGN's owner, the Tribune Co., may try other national teletext ideas.

There are currently about 35 television services delivered nationally via satellite to receiving stations at cable system headends or studios, and all are potential users of the teletext technique for slipping additional information into the already distributed television signal. The teletext data may or may not be related to the video/audio signal. So far we have talked primarily about services where there is no connection, but an example of a service where there is a relationship is the Weather Channel, a new venture of Landmark Communications. The Weather Channel provides national, regional, and local forecasts and will carry advertising as well. While part of the service will be the normal video signal showing meteorologists and weather maps, another part will be digital information carried in the vertical blanking interval of that signal. Receiving cable systems will use a Landmark-developed unit that decodes the data and generates a display. Because the receiving units are addressable, subscribing cable systems can receive, via teletext, tailored information for display at predetermined times. In addition, in the evening when the video part of the Weather Channel is preempted by pay TV for a few hours, the weather service can continue using the teletext-delivered data.

Of course, all of this is only a version of the teletext technique, because the ultimate user in the home has no ability to select pages, but the satellite delivery of teletext to cable system studios or headends is certainly bringing closer the day of widespread teletext to the home. (It should be noted that if a cable system simply passes along a teletext-bearing television signal without erasing the data in the vertical blanking interval, a home viewer with the proper decoder can then receive true teletext; but for various reasons cable systems may decide not to allow teletext signals to pass through untouched.) Some satellite-delivered services have already begun preparations for mixing teletext destined for the cable operator with teletext destined for the cable subscriber. One example is Time, Inc., owner of the Home Box Office pay TV service and ATC, a cable system operator. (Time, Inc.'s teletext ventures are discussed in the following section on teletext as part of the local cable system.)

Teletext on Cable

On cable systems themselves, the teletext technique has been used in a number of ways, from the simple data-delivery mode to full-channel teletext. The immediate advantage for cable systems is that existing channels already in use can be utilized to carry additional services if the data are multiplexed into the vertical blanking interval. Some cable operators have taken traditional teletext systems and adapted them for use on one or two of the cable's channels, while other cable companies have sought to provide a completely integrated cable transmission system that would use a mixture of established data communications techniques, local area network techniques, polling, and teletext techniques, so that the most appropriate technology is used for each part of the total package of cable-carried services.

At the less complicated end of the scale are the cable systems using teletext in exactly the same manner as a broadcast television station. A teletext system is installed to enhance one channel (or perhaps a few channels), and the home viewer needs a teletext-equipped television set, just as if the enhanced channel were being broadcast over the air. In Danbury, Connecticut, Dow Jones has tested a service that did just that. The channel used for teletext was already a billboard channel, carrying pages created by the Danbury *News-Times,* owned by the Ottaway Newspaper Group, a Dow Jones subsidiary. Approximately 50 homes were given teletext-equipped television sets in order to receive the teletext portion of the newspaper channel. The billboard part of the channel, delivered on the cable as a normal video, is created by editors at the *News-Times* using a character generator.

For the pages delivered in the teletext mode, a 100-page magazine was created containing about 80 pages written by the *News-Times* and 18 pages provided by Dow Jones. The technical system used was the French Antiope/Didon system; funding for the trial was provided jointly by Dow Jones and Antiope Videotex Systems.

A similar trial is underway in Louisville, Kentucky, conducted by the Dissly Research Corporation, a subsidiary of the Louisville *Courier-Journal* and Louisville Times Company. As in Danbury, the project is a joint effort with Antiope Videotex Systems. Cable channels will be leased for the trial and will carry rolling pages of news and other information as the normal television signal, while the teletext portion of the channel will carry additional newspaper items including classified ads. During the first phase of the project, in early 1982, ten teletext-equipped television sets were placed in homes to determine the kinds of information most appealing and most suited for teletext delivery (i.e., information that benefits from the viewer's ability to select pages).

Reuters, a company that has been involved longer in cable television, has also developed a system that uses more than merely the vertical blanking interval. In the mid-1970s, Reuters began using cable television systems to distribute business information to corporate clients in New York City [4]. The cable system provided the appearance of interactive service while still using a one-way cable because the teletext technique was employed—an entire file of pages was transmitted in a continuing cycle. But because a full televisionwide channel was being used, the entire file could be transmitted within a few seconds, thus reducing the average waiting time to about one second. This system has evolved into a full-channel teletext system available not only in New York but also elsewhere in the United States, delivered via satellite. The eventual system used was developed by IDR, Inc., a subsidiary of Reuters.

Reuters does not, however, use the teletext terminology of magazines and cycles, but rather the data communications terminology of packets and data streams. The data stream is divided into 1,000 logical groups, with each group able to contain up to 1,000 screens of text. Physically, the data stream is transmitted in packets that are transparent to, and unrelated to, the logical structure. The terminals used by Reuters's teletext customers can display either 32 or 64 characters on a line and 16 lines on a screen or page. Simple graphics are also possible, and the display can be in eight colors.

The full-channel nature of the Reuters service means that instead of 100 pages being transmitted in 20 to 25 seconds, over 10,000 pages can be transmitted in the same amount of time. The originating computer can decide to cycle some items more often than others, so the complete cycle for certain data groups may vary from one to eight seconds. If a subscriber is viewing a given page or screen of data, that screen will change to reflect updates to that page as the page is transmitted in successive cycles. The types of information distributed by Reuters in this way include stock and commodity quotations, foreign exchange data, money market information, and financial news.

In 1982 Reuters announced that it would begin marketing this service to a wider audience rather than to specialists by offering a new, less expensive terminal. The terminal is designed to be used with cable television and is specifically not to be technically compatible with other teletext techniques.

Naturally, full-channel teletext prevents a channel from being used as a normal video channel, but in the cable television environment spare channels sometimes exist. Because the digital data are still multiplexed into a television signal, albeit no lines are left for video, that signal can be more or less handled by the equipment throughout the cable system as just another television channel. Moreover, the one-way signal behaves like a two-way or interactive service in the eyes of the user. Thus full-channel teletext seems to be an important technique in the cable environment (and to a much lesser extent in other private television systems such as MDS and STV).

One company that has begun to explore both full-channel teletext and vertical interval teletext is Time, Inc. Time changed the revenue situation of the cable television industry in the late 1970s with the highly successful pay television service, Home Box Office. Cable subscribers were easily convinced to pay additional monthly charges in order to receive the movies and other entertainment programs delivered uncut by Home Box Office. Time now wants its new teletext service to be the Home Box Office of the teletext industry. With that goal in mind, Time purchased a Telidon system in early 1981 and created an editorial staff initially numbering about 30 to begin the service. Reportedly, the objective is to produce 12 to 25 different teletext magazines—Time does not necessarily use the term "magazine" though—of about 100 to 150 pages each.

From the fall of 1981 to the summer of 1982, Time tested its teletext ideas at its Consumer Communications Center in the basement of the Time-Life Building in New York. Approximately 1,000 people (paid volunteers) were given the opportunity to view pages of text and graphics and were then questioned about their reactions and opinions regarding the pages. In addition, the volunteers were observed through one-way mirrors by researchers looking for more subtle reactions. Some of the preliminary conclusions reached by the Time researchers were that teletext pages have to be entertaining and useful, and it helped if there was a high degree of interactivity with the teletext pages. Time has established a subsidiary, Time Video Information Services, Inc., to develop the teletext service that will undergo field tests during 1983. When the service eventually begins, it is expected to carry advertising, to rely on local newspaper organizations to provide the local interest pages, and to be so up-to-date that no item would stay on the service for more than four hours without being replaced or rewritten (see Figure 4.3).

The field testing is to take place on two cable television systems owned by ATC, a Time subsidiary; the systems are in San Diego, California, and Orlando, Florida. Approximately 150 homes in each system will be given equipment to receive and decode the teletext pages and will be able to use the service free of charge.

It will no doubt be several years, at least, before a nationally distributed teletext service affects the cable industry in the way that pay television did. As long as teletext

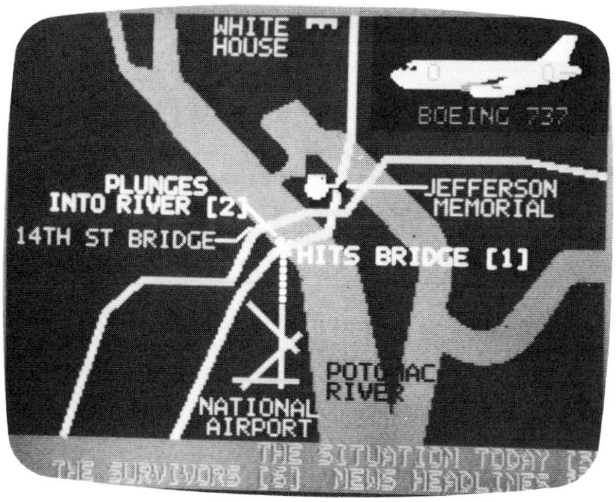

Figure 4.3. Sample page from Time's teletext service. (*Courtesy of Time Video Information Services.*)

is primarily used for text (i.e., for news and information purposes), there is much less chance that consumers will pay as much for the service as for entertainment. However, major publishing and media conglomerates like Time, Inc., have begun the process of including large-scale teletext services in their bag of businesses.

On the equipment side of the cable environment, equipment manufacturers have recognized the advisability of building the teletext technique into integrated systems for both traditional and nontraditional cable services. A good example of such an integrated system is that produced by Tocom, Inc., one of the companies involved in two-way cable television experiments in the early 1970s and a developer of security and alarm systems designed for cable television.

The Tocom 55 Plus system is actually a group of products designed to permit cable operators to move from technically simple ancillary services to more complex services when economically feasible [5]. For example, there are two types of set-top Tocom 55 Plus converters, one without teletext capability, and one with that capability, which can be substituted for the less expensive converter as necessary. Moreover, the teletext-equipped converter can include as an option an interactive data communications facility. (See Table 4.3 for a list of the Tocom converter features.)

At the less complex and less expensive end, the Tocom system provides for 55 channels of television, 32 classes of service, pay TV channels and secure scrambling of those channels, codes for parental control of types of programs and other forms of restricted access, and an emergency alert. The service classes are a means of coding programs so that a single channel can carry several different classes of programs, and only the subscriber who pays for certain classes is able to view the programs in that class. Another way to attach subscriber charges to programs is use of subscription programming codes, which permit any viewer to be authorized for up to four subscription services—even if the service is actually teletext—although the system itself can carry more than four subscription services. Another form of control over programs is the parental discretion feature, which allows programs to be tagged with up to eight hier-

Table 4.3. Features of the Tocom Terminals in the Tocom 55 Plus System

Feature	5504A Terminal	5510A Terminal
Baseband addressable	X	X
Video channels	X (55)	X (55)
Vertical interval teletext		X (55)
Full-channel teletext		X
Service classes	X (32)	X (32)
Subscription programs per sub	X (4)	X (4)
Scrambling	X	X
Emergency alert	X	X
Parental discretion coding	X	X
Access code	X	X
On-screen channel number	X	X
On-screen clock	X	X
Remote control	X	X
VTR output	X	X
Interactive data retrieval		optional
Formated data entry		optional
Pay-per-view		optional
Opinion polling		optional
Channel monitoring		optional

Source: Tocom 55 Plus General Information Manual, 1981, p. 11.

archical codes; a program with a code exceeding that selected initially by the subscriber can be viewed only if the subscriber types in an eight-digit access code. The emergency alert feature sounds something like a Big Brother nightmare—the cable operator can cause a home television set to turn itself on, automatically tune to a certain channel, turn up the volume, and deliver an emergency message. (Subscribers can elect not to have this feature.)

The more advanced Tocom unit incorporates teletext as well as the polling technique and interactive data communications in order to handle digital signals for a number of services. Vertical interval teletext is possible on all 55 channels using lines 17 and 18 for transmission of pages. These pages currently have a display format of 16 rows of up to 32 characters each. An example of a Tocom teletext page is shown in Figure 4.4.

The teletext part of the Tocom system in the vertical interval mode is somewhat more limited than the teletext systems used by broadcast television stations because the number of characters per row is fewer and the data rate currently used is slower. Thus in the Tocom system, a magazine of 100 pages takes approximately 100 seconds (or one-and-two-thirds minutes) to complete a transmission cycle, while a similar magazine on the broadcast systems now in use can be transmitted in about 25 seconds. In a move to alleviate the potentially long average waiting time, Tocom has suggested that pages can be grouped into categories where the order of the pages within a category is unimportant, and the viewer selects categories, not pages. Consequently, if there are ten pages in a category evenly distributed throughout a cycle of 100 pages, a viewer would never wait more than ten seconds to see a page from that category, and the mean waiting time would be five seconds. As in other teletext systems, of course, index pages can be inserted into the cycle more often so that there is virtually no waiting time for indexes. Tocom argues that vertical interval teletext saves valuable channel space now devoted to the billboard channels, where viewers have no ability at all to select pages or even categories of pages.

Figure 4.4. Sample teletext page on a Tocom cable system. (*Photograph provided by Tocom, Inc., Dallas, Texas.*)

The Tocom system also includes full-channel teletext, with up to 200 scan lines (out of a maximum of 262.5, both fields) used to carry multiplexed data. The Tocom scheme calls for the 200 scan lines to be allocated to different types of pages with different cycle times. The first 40 scan lines carry index pages such that retrieval of an index page would be virtually instantaneous. The next 60 pages carry "rapid access" information such as financial news or stock reports, and the arrangement of the data will give a maximum cycle time of ten seconds. The bulk of the scan lines, from 100 to 200, carry categorized information such as classified ads, which could have a somewhat longer maximum cycle time.

In addition to the teletext capabilities, the Tocom system includes the polling technique and interactive data communications. For opinion gathering and simple subscriber responses, the home terminal can accept any two-digit number. For pay-per-program purposes, a subscriber can type in an access code in order to be billed for a specific program. For statistical purposes, the polling mechanism can accumulate aggregate figures for the hours of viewing per channel for the entire system (Tocom says that individual records of viewing are not collected in this procedure, and that individuals can have their television excluded from this channel monitoring if desired). For the more general case of initiating a digital communication from the home, a keyboard can be added to transmit data on one of three set frequencies so that home terminals share the response frequencies in both a time division and frequency division manner.

Controlling all these forms of digital signals are a collection of processors at the headend, which can be added one by one as higher levels of service are implemented. A video processor, for example, is needed for each of the channels that are to be subject to control, such as pay TV channels. This processor is also used to insert data into the vertical blanking interval of that channel. A second type of processor is used for the full-channel teletext service. A third type of processor handles the interactive services, accepting data and routing the digital messages to external computers. Overseeing the entire system is a program control processor.

Perhaps in line with Tocom's integrated system, certain types of services are seen as more suitable for one delivery technique than for another. For the sake of teletext, it is

Table 4.4. Tocom's List of Advantages for Teletext on Cable Systems

1. Headend costs for teletext are relatively low.
2. Headend costs are completely independent of the number of subscribers using a teletext service.
3. The technology puts no limit on the number of simultaneous users.
4. Access time is not affected by the number of simultaneous users.
5. Teletext can be used on one-way cable systems.

Adapted from: Tocom 55 Plus General Information Manual, 1981, p. 22.

interesting to note that Tocom has suggested that up to 95 percent of all data retrieval needs in the home can be accommodated with teletext, which is essentially a one-way transmission, without the complexity and expense of a fully interactive system where signals must be returned to a cable headend computer or routed through to a distant computer. Tocom's belief in teletext stems from the relatively low cost of teletext equipment and the fact that the number of simultaneous users has absolutely no effect on the speed of delivery of the pages. The advantages that Tocom sees for teletext are listed in Table 4.4. In a fully interactive system, on the other hand, a more expensive computer system may be needed to handle the processing of incoming transmissions, and the retrieval time is affected by the number of simultaneous requests or messages to be processed, while at the same time the number of simultaneous users is also restricted by the channel space allocated to return signals and the methods used to divide space and time.

Another company that has voiced a belief in the advantages of teletext on cable is the Jerrold Division of General Instruments Corporation, one of the established leaders in the cable equipment industry. The Jerrold group of cable subscriber home terminals also exhibits a stepped approach to adding the more complex services. At one end are the security and pay-per-program features using polling techniques. The pay-per-program system itself can be upgraded from a system that requires the subscriber to telephone the cable operator to receive an authorization number, to a system where ordering can be done directly through the home terminal.

At the higher level of service, the Jerrold system includes vertical interval teletext and full-channel teletext. A Jerrold brochure suggests that teletext techniques can deliver a thousand or more pages of news, weather, stock reports, and the like, with extremely short response times. (The maximum cycle time, of course, is directly affected by the total number of pages in the cycle, and a full-channel teletext service with only a thousand pages in the cycle could have a maximum cycle time of only two or three seconds, or a mean waiting time of a second and a half.)

Beyond that, Jerrold is beginning to offer a fully interactive data communications system based on local area network techniques. (In 1981, General Instruments purchased Sytek Corporation, a company specializing in broadband local area networks.) In structuring the complete package, a Jerrold official has outlined how different transmission techniques can be used for such services as TV games, security, pay TV, information delivery, and home shopping (see Table 4.5) [6].

It is likely that as systems such as Jerrold's are developed, the teletext and interactive data subsystems will be physically and functionally separate. One model of set-top converter, for example, will be used to receive and decode teletext data, whether vertical interval or full channel, and a separate piece of equipment, or set-top unit, will be added for accommodating the transmission of data signals, perhaps using a full alphanumeric keyboard in the home or office. This separation follows from the belief that a true

Table 4.5. Possible Procedures for Providing Different Services via Cable

Application	Procedure
TV games	One-way signal with a 200-kHz bandwidth
Pay television	Two-way signal with a 200-kHz bandwidth
Alarm monitoring	Two-way signal with a 200-kHz bandwidth
Information delivery	One-way teletext on a TV channel of 6 MHz
Home shopping	Two-way signal with a 100-kHz bandwidth

Based on Thomas E. O'Brien, Jr., A Unified Approach to Data Transmission over CATV Networks, *Cable '81 Technical Papers* (Washington, D.C.: National Cable Television Association, 1981), p. 123.

interactive data communication is essentially a different activity than merely looking at pages of text on a casual basis while watching television. In fact, the terminal used for the highly interactive data communication may not be the television set at all but the home microcomputer, and therefore there is no need to give the subscriber a single unit that is complex enough to handle all activities but that is more probably used for only one set of activities in a given location.

Other manufacturers have also begun the process of adding teletext capability to home converters, and the process is only somewhat hampered by the lack of national teletext technical standards. Unlike broadcast television, the cable system is something of a closed environment. Typically, the cable operator rents or leases the set-top converter to subscribers, and subscribers are not permitted to take the converters to other locations. Therefore the cable operator can choose to design and construct the equipment in any way that seems feasible. But the environment is not quite that closed, and cable operators and equipment manufacturers rely on the availability and price of components that are indeed affected by what other companies do elsewhere. For example, Jerrold's parent company, General Instruments, has a division manufacturing the computer chips for U.K. teletext and videotex systems, Jerrold has elected in the United States to build systems using the CBS/AT&T/Telidon/Antiope "standard" for transmission and coding in the belief that this standard will prevail here.

Despite all this activity in using the teletext technique on cable, some people have suggested that the broadcast teletext procedure cannot be adopted in full for cable applications. The problem, they say, is that cable systems will not be able to handle the high instantaneous data rate associated with broadcast teletext (e.g., 5.7272 megabits per second). It was feared that this data rate would cause interference with other signals on the cable, or that the amplifiers and other pieces of equipment that are part of a cable system would not be able to process the data signal without corrupting the data. So in late 1981 and early 1982, a committee of the National Cable Television Association ran some tests of teletext data rates and pulse shapes. The first test took place on a Viacom cable system on Long Island, N.Y., and a second test was conducted nationally using the satellite facilities of Time, Inc.'s Home Box Office and a Time/ATC cable system in San Diego. The results were formally inconclusive, although there seemed to be ample evidence that there was no cause for alarm [7]. The high data rate of 5.7272 megabits per second did not seem to result in either interference or self-corruption, and the previously suggested pulse shape, a "100 percent cosine rolloff," did perform better than alternative pulse shapes.

Other Systems

In the early 1980s, as basic cable systems began adding the newer security and pay television techniques, a number of services were introduced that used teletext or teletextlike procedures to deliver information or entertainment to subscribers. These types of services may indeed bridge the gap between traditional cable service and the advanced services provided by teletext (and, beyond that, by interactive data communications on cable).

The prime example is Playcable, a joint venture of Mattel Electronics and the Jerrold Division of General Instruments. The Playcable concept is a little complicated in that it is based on equipment that can be a video game machine used alone, or a video game machine that requires connection to a cable television system with complementary equipment at the cable headend, or can be a stand-alone microcomputer equal to or better than, say, an Apple or a TRS–80. The equipment is physically a set of units called the master component, the keyboard component, and the cable adapter; to be complete, this set of units requires the use of a television set as the display device and possibly plug-in cartridges if the equipment is not connected to a cable system. To add to the tangle, some of the equipment is marketed by Mattel (Mattel's Intellivision), and some of the equipment is marketed by Playcable (directly to cable operators rather than to consumers). And some parts of this package were introduced to the marketplace before others.

Mattel's Intellivision was first introduced with only the master component, containing the video games feature, apparently in the hope that the public would buy video games much more easily than microcomputers. The games themselves are sold in cartridges that plug into the master component. During 1982, Mattel began selling the keyboard component with the microcomputer capability when attached to the master component. A customer can buy the video games component alone and later add the microcomputer equipment and resulting processing power that permits even more cartridged applications to be used.

Playcable, the service designed specifically for cable television, requires the Intellivision master component in the home and a cable adapter. The games and other Playcable programs are fed from the cable headend to the home, presumably giving the consumer access to a far greater library of applications programs than would ever be purchased as cartridges to be kept at home (see Figure 4.5). In addition, the Playcable subscriber can have access to the advanced packaged programs if the keyboard component is also purchased. Such advanced programs include educational packages such as language courses and speed reading training, and analysis programs such as one to compute personal investment possibilities.

Germane to teletext development is the fact that Playcable uses the teletext principle of cycles of data delivered in a one-way system in order to make the games and other programs appear interactive. The data are not multiplexed into a vertical blanking interval, although this procedure was tested, but instead are transmitted as a narrowband data stream in the FM radio portion of the cable spectrum. One of the reasons that Jerrold chose to put the data stream in its own narrowband subchannel is that the steady data stream can operate at a slower rate than the high, but discontinuous, burst data rate

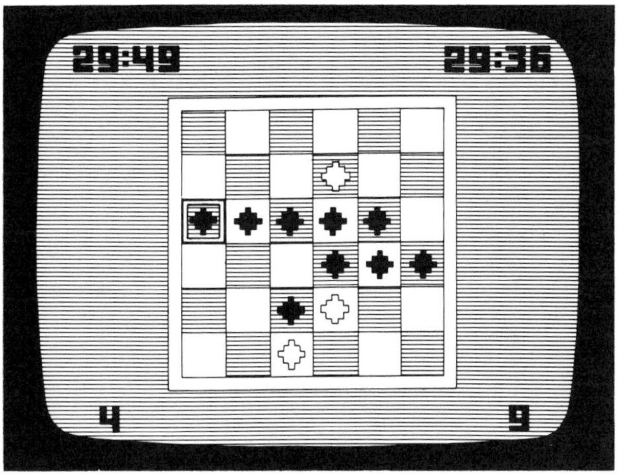

Figure 4.5. Sample Playcable screen. (*Courtesy of Playcable.*)

of teletext. The decoding equipment used for the slower data rate is less expensive [8]. The actual flow of information to the television screen, however, is only slightly slower in the Playcable approach than in a teletext approach, even though the data rate used for Playcable is only 14 kilobits per second, while the burst data rate for most of the teletext systems in use in the United States is about 5.7 megabits per second.

In order for a subscriber to receive Playcable, additional equipment must be in place at the cable system's headend or central control point. This headend system is built around a Digital Equipment Corporation minicomputer, the PDP 11/03. Games and other programs are stored on floppy diskettes, which are placed on the minicomputer's disk drives as easily as records on a phonograph. The data are then fed to a data channel card, with a separate card required for each data channel. (Data channels can be added, with 200-kHz spacing, as space permits in the cable's usable spectrum.) From the channel cards the data are passed through an rf (radio frequency) data modulator, or modem. The actual placing of the data channels does not require that the channels be contiguous in one frequency area but can be interspersed among other FM channels. One data channel carries a directory or menu of available games and programs, and other data channels carry the data for those games and programs. The home terminal automatically knows which data channel to tune to after the subscriber makes a selection from the directory listing. For some applications, the desired game or program then selected by the terminal is received and stored in an 8,000-byte memory (using 10-bit bytes) at the terminal available for local processing. For other applications where the program requires more than 8,000 bytes, the additional data are picked out of the continual cycle by the terminal as necessary.

As with teletext, the subscriber can experience a waiting period following the selection of an item before the item appears, because a single data channel may be carrying up to 21 games or programs in its cycle. The average waiting time is about ten seconds.

During 1980 and 1981, Playcable was test marketed on five cable systems: in Minnesota, California, Idaho, Illinois, and Mississippi. For each system, 200 cable adapters were built, and all were eventually used except on one new system that was just being built. Subscribers had to buy the Mattel master component either from the cable operator or from any store selling the unit, and they had to pay for the service. In mid-1981, Playcable concluded that, with some adjustments in the marketing, the Playcable service would be successful, and began selling the system/service to cable television operators nationally.

Another example of a service adapted for the cable television environment using teletext concepts is SourceCable. The Source is a large data base of news and business information, consumer information, analysis programs, games, and the like used by people with microcomputers (or computer terminals) and a telephone. With the telephone network to connect the microcomputer or terminal to The Source computers, usually by a low-cost data network, users cannot only interact with the large data base but can also send electronic mail, process text, shop electronically, and accomplish a host of other tasks made possible by accessing the resources of a large computer.

For the cable environment, The Source developed SourceCable, a selection of information from The Source with a page format (see Figure 4.6). These pages are stored at a cable headend and delivered in a one-way cycle using vertical interval teletext, full-channel teletext, or a data transmission method such as that developed for Playcable. In order to view the pages and to appear to be interacting with a data base, the home subscriber needs a decoder and a keypad. At any given time, the store of pages at the headend may total approximately 2,000 pages, with updates to pages or replacement pages received either continually or periodically by telephone lines from The Source computers. The actual number of pages in the cycle within the cable system depends, of course, upon the delivery technique and the maximum waiting time that the cable operator wishes to inflict upon the user. Information packaged into the SourceCable cycle includes news, weather, sports, energy hints, selections from Reader's Digest publications (Reader's Digest purchased The Source in 1980), health information, a list of toll-free telephone numbers, movie reviews, culinary advice, plant and gardening advice, and even Reader's Digest humor columns.

Figure 4.6. Sample page from SourceCable. (*Courtesy of the Source Telecomputing Corporation.*)

Table 4.6. Use of the Teletext Technique by Major Manufacturers of Cable Television Home Converters

Manufacturer	Teletext Option
Jerrold	Yes
Oak	Possibly
Pioneer	Not available
Scientific-Atlanta	Not available
Tocom	Yes
Zenith	Yes

The SourceCable service was created in 1981 and initially developed as part of Cox Cable's trial of its own Index system for one-way and two-way information services on cable.

The result of all this activity is that the teletext technique seems to have a promising future in cable television as a way to increase the capacity of the cable and to provide the appearance of interaction on essentially one-way cable systems. A number of cable television equipment manufacturers are including the teletext technique in integrated systems composed of several data delivery techniques (see Table 4.6).

In the overall view, however, cable is still a much smaller market than broadcast television. At the end of the 1970s, approximately three-fourths of all television homes were *not* subscribing to a cable television service. And of the two dozen largest television markets in the country, only five were served by cable television systems. The major television markets will probably not be fully served by cable television before the end of the 1980s.

Moreover, the most likely new additions to cable services are security and pay-per-program features, which seem most suited to polling techniques to accomplish the digital control. An early 1981 list of enhanced cable systems contained a little over 35 cities where a security alarm monitoring service was provided by the cable television system, while only a handful of systems included, or were planning to include, other services using teletext or interactive data communications [9]. As mentioned above, the biggest concern for cable operators is to introduce addressability into their systems to facilitate the alarm monitoring and channel monitoring services (e.g., for authorizing premium programming on a pay-per-program basis). Addressable converters for the cable subscriber were first introduced as a commercial offering in 1979 by Oak Communications, and a cable industry newsletter speculates that 80 percent of cable subscribers could be using addressable converters by 1990, although questions of cost and the appropriate technology (and the value of the resulting services thus enabled) still hamper the spread of these converters [10].

The cable environment, therefore, is a fertile ground for teletext, but teletext will have to take its place among a growing variety of other data handling techniques.

References

1. National Cable Television Association, Media Services and Research Department, Enhanced Services List, March 26, 1981.
2. For a history of two-way cable television, see Richard H. Veith, *Talk-Back TV: Two-Way Cable Television,* TAB Books, Blue Ridge Summit, Pa., 1976.

3. See, for example, Margaret Yao, Two-Way Cable TV Disappoints Viewers in Columbus, Ohio, As Programming Lags, *Wall Street Journal,* September 30, 1981; David Burnham, The Twists in Two-Way Cable, *Channels,* June-July 1981, pp. 38–44.
4. Details are contained in Michael J. Reilly, Generating Teletext Characters by Row-Grabbing Technique, *CED,* December 1981, pp. 20–22.
5. The description of the Tocom system is based on Tocom's General Information Manual, 1981.
6. O'Brien, Thomas E., Jr., A Unified Approach to Data Transmission over CATV Networks, *Cable '81 Technical Papers,* National Cable Television Association, Washington, D.C., 1981, pp. 119–123.
7. Lopinto, John, NCTA Teletext Testing Program, speech at Viewtext 82, New York, April 13–15, 1982.
8. Dages, Charles L., Playcable: A Technological Alternative for Information Services, *IEEE Transactions on Consumer Electronics* CE–26 (3): 482–486 (August 1980). See also Susan Spillman, Playcable: Not Toying Around, *Cablevision,* May 25, 1981.
9. National Cable Television Association, Media Services and Research Department, Enhanced Services List, March 26, 1981.
10. Kagan, Paul, *Cable TV Technology* (newsletter), April 23, 1981.

5

Formats and Standards

As teletext and videotex systems were developed, debates over formats, features, and standards surfaced continually. The future for teletext and videotex—systems designed for the mass market—seemed to hinge on national and international agreements regarding some basic level of operation, so that equipment could be manufactured on a large scale by a variety of companies. In England in the early days of teletext, a committee of representatives from the television industry, television set manufacturers, and the semiconductor industry was able to agree on a set of coding and transmission standards that enabled teletext to grow. In France, Canada, Japan, and elsewhere, similar agreements have been reached that present a consensus in each country, although certainly not a consensus internationally.

In the United States, the existing computer-based information services began evolving their own teletext and videotex systems while at the same time evaluating, and being romanced by, individual foreign systems. On the videotex side (i.e, systems using telephone lines), decisions regarding the transmission and display formats could be made by single organizations because a basic level of standardization had already been achieved for data communications via telecommunications networks. That is, virtually all videotex systems use the ASCII code (American Standard Code for Information Interchange) for the digital representation of characters, numbers, and common punctuation, and all systems similarly use standard practices for passing the data through modems (*mo*dulator/*dem*odulator) to, and through, the telephone network. Nonetheless, arguments regarding standards for organizing the data stream and for presenting the screen format or display have dogged development because of the certain knowledge that standardization of these elements would permit economies of scale and drive down costs.

For broadcast teletext and to a lesser extent for teletext in general, all the same arguments apply, with an additional concern about how to multiplex the digital information into the television signal. Previously, the only use of the vertical blanking interval had been for test and reference signals and the line 21 closed captioning system as it developed. Because the line 21 system and the teletext systems in England, France, and Japan all developed dissimilar procedures for incorporating data into a television scan line, additional arguments arose in the United States over the relative merits of these teletext transmission schemes.

There did, however, seem to be a solution for the case of broadcast teletext. There was, in fact, a final arbiter who has jurisdiction over standards for broadcast signals: the Federal Communications Commission (FCC). It was the FCC that judged the arguments preceding the introduction of the line 21 system and permitted that system to develop. It was the FCC that granted special experimental licenses to the broadcasters who began trial teletext services in the late 1970s and early 1980s. And it was the FCC that began accepting arguments from interested parties in favor of one teletext standard or another.

But in late 1981, within the general deregulatory mood in Washington, the FCC initiated a rule-making procedure in which the commission stated that it did not intend to impose any specific coding and transmission standards, and would even open up line 21 for general teletext use. As this is being written, the FCC has yet to resolve the issue but will probably endorse a marketplace decision. In any event, it is likely that in the years ahead, the formats and standards for coding and transmission will continue to be a concern, not only because broadcast stations will be looking for means of enhancing their system within a given standard, but also because some organizations, such as cable television companies and MDS and STV operators, are able to act without the need for FCC approval in this regard. Additionally, the videotex systems that succeed will bring their own weight to market forces favoring certain display formats.

Many of the disparities and dissimilarities in display and transmission procedures have been mentioned explicitly or implicitly in earlier chapters. There is the British standard, the French standard, the Canadian display standard, AT&T's own display standard, the CBS/French transmission standard for broadcast signals, joint standards in Europe and another set of joint standards in North America, and an accumulation of individual systems among telephone-based services and within the cable television environment. It is the purpose of this chapter to try to bring together these concerns in a more detailed presentation, with particular emphasis on teletext.

Page Format

Computerized information systems originally described by the terms "teletext" and "videotex" were, by definition, designed for use with television sets. The information in the form of characters and graphics was to be displayed on raster scan television sets. Because television sets use a display technique basically different from the normal computer terminal (i.e., a screen display on a television set is composed of a series of horizontal scan lines of varying intensity), one of the first characteristics of teletext/videotex systems to emerge was information displayed on static screens where the number of characters on a row and rows on a screen was affected to a great extent by the number of scan lines in current television systems.

Actually, there is a complete list of factors that affect the optimum number of rows and characters, including such items as viewers' average distance from the screen, whether lower case letters are used, whether letters with diacritics are used, whether the scan lines of alternate fields contain the same or a slightly different pattern, whether the display is in black and white or color, and so on (see Table 5.1). But as Figure 5.1 shows, a minimum number of dots is required to produce legible text, and characters need to exhibit common proportions. Additionally, blank areas are required between rows and characters. Thus the actual number of scan lines can place a definite limit on the total number of characters per screen. There is also a problem in television called "overscan," the tendency of television sets to be imperfectly aligned with respect to the total amount of picture, from edge to edge, seen by the viewer. Television camera operators know that approximately 90 percent of the picture that they see may actually be viewable on the home television set, and titles or other text are kept to the inner 80 percent of the screen just to be safe. In other words, a camera operator would not focus on a line of text in which the letters appear at an edge of the camera's monitor, because almost certainly the home viewer would not be able to see the letters near the edge. Similarly, teletext displays must not use every possible scan line or every possible dot within a scan line, out to the edges, so that text will not be out of sight at the sides, top, or bottom.

Given these limitations and the fact that scan lines from alternate fields are "laid down" on top of each other when using character-generating circuits in the television set in a noninterlaced mode (i.e., the total number of lines available equals one-half the number of scan lines in a frame), British engineers working on the first teletext system concluded that the maximum number of characters and rows that could be displayed in a single frame was 24 rows of 40 characters. From British television's 625 scan lines per frame, subtracting the lines in the vertical blanking interval leaves approximately 280 lines (in each field) for character display. Assuming ten lines are needed to produce a legible character and the space around it, 240 lines are used for the 24 rows, which is within the safe viewing area for broadcast television. Based on assumptions about the number of lines needed to produce a readable character, the reasonable proportions of a character and a reasonable viewing distance, and considering the technical limitations of the television display, the 24 × 40 page format for teletext was standardized. (Another standard feature agreed upon for teletext in England was a switch in the viewer's

Table 5.1. Page Format Considerations

1. Type of material to be viewed.
2. Distance between viewer and the television screen.
3. Whether teletext is to be superimposed over normal video.
4. Resolution limitations of the human eye.
5. Resolution limitations of the television screen.
6. Whether set-top decoders are used.
7. Timing limitations when accessing digital memories.
8. Ability to display non-English characters.
9. Readability of the displayed page.
10. Aesthetic value of the character set.
11. Similarity with the formats of other computerized information systems (e.g., videotex).
12. Cost.

Based on: Stuart Lipoff, John Lopinto, and Robert Siedel, Final Report of the Ad Hoc Page Format Working Group, *BTS/Teletext Subcommittee, Interim Report,* vol. II, Electronic Industries Association, Washington, D.C., 1981, p. 126.

2 x 3 character
in a 3 x 4 matrix

4 x 5 character
in a 5 x 6 matrix

5 x 7 character
in a 6 x 8 matrix

5 x 7 character
in a 6 x 10 matrix

7 x 9 character
in an 8 x 12 matrix

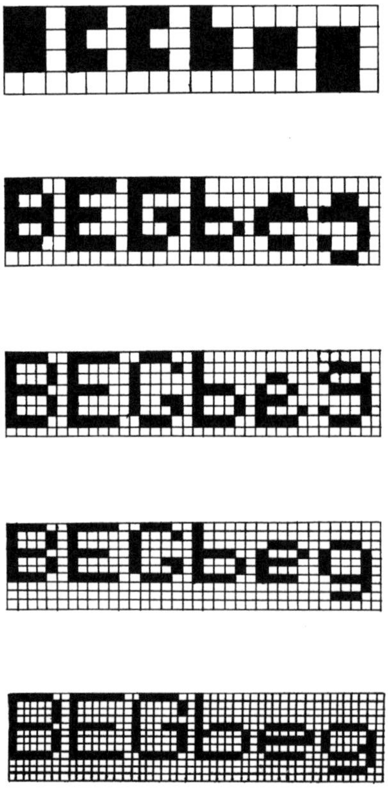

Figure 5.1. Examples of various character formats.

remote control unit to automatically double the height of every character so that either the top or bottom half of a screen could be viewed in double height, under control of the viewer, to insure that text on a television screen could still be read across a room.)

In the United States, even prior to the development of teletext, various companies had grappled with the problems of creating text on television using character-generating circuits in the sets, but with a television system that has only 525 scan lines per frame, not 625 scan lines. The solutions chosen reflected not only the fewer number of scan lines but also different sets of assumptions about who the average viewer would be and how far from the set the viewer would sit, and how crude a character set the viewer would accept. Thus some systems display only capital letters, and others display lower case letters but letters with descenders (e.g., g or y) are "bumped up" above the line of other lower case letters, and still others display fewer rows of characters per screen than alternate systems. It is almost impossible to say that a consensus developed, although a number of manufacturers did begin using a display format of 16 rows of 32 characters because an integrated circuit, or computer chip, was produced for that format and has been widely available. The Radio Shack videotex system, for example, is based on a display or page format of 16 × 32; the line 21 captioning/text system displays 32 characters on a maximum of 15 rows per screen; many character generators for cable television display 32 characters on either 8 or 16 or 20 rows; and the early teletext experiments of KSL–TV used a page format of 32 characters on up to 20 rows.

When the subject of page or screen formats for teletext began to be debated as such in the United States, the immediate question was whether or not the European format of 24 × 40 (or even 25 × 40) could be adopted without change, or whether the lack of 100 scan lines would necessitate a page format specifically for NTSC 525-line television. Moreover, the possibility that the development of teletext would require the use of set-top decoders, which generate a modulated signal to be fed into the antenna terminals on the television set, put additional constraints upon choosing the optimum page format. (The available bandwidth directly affects the possible number of characters per row, and assuming an available bandwidth of 3 MHz when connecting a set-top decoder to antenna terminals, 40 characters will not fit on a row if a 7 × 9 dot character matrix is used, but 40 characters *will* fit if a 5 × 7 dot matrix is used [1].)

The prevailing opinion as this is being written is that 40 characters per row is quite possible and even advisable within the NTSC television system. In the future, new technologies and better equipment will almost certainly expand that number to perhaps 80 characters or beyond. But in the foreseeable future the technical limit will be 40 characters per row. The arguments favoring 40 characters over 32 characters per row are essentially on the order of ''more is better''. If 40 characters can be decently displayed, why limit the creative and productive aspects of teletext page origination by accepting a limit of 32 characters per row? An additional incentive to accept 40 characters is the fact that Western Europe has agreed upon 40 characters per row, and therefore teletext (and videotex) data bases can be used internationally without changing row formats.

The matter of the optimum number of rows per screen is less easily solved. The lack of 100 scan lines, in comparison with European television, suggests that 24 rows is not practical on 525-line television. In fact, a 1979 paper by one of the leading British teletext engineers suggested that, in the United States, 20 rows would be the preferred format [2]. This would be the maximum number of rows possible assuming 10 lines per character display and staying within the safe viewing area.

However, given both the desire to maximize the number of rows and to deviate as little as possible from the European systems, ways were developed to display 24 rows of readable characters on U.S. television sets. Zenith Radio Corporation, for example, has argued that there are indeed several ways to reasonably create the 24 rows [3]. One of the cleverer methods used by Zenith is called raster compression—the picture is reduced slightly in the vertical direction so that 480 scan lines are definitely viewable, thus allowing ten visible lines to define each row. This compression happens automatically when the viewer selects the teletext mode, and often it is not even noticeable unless a knowledgeable person points it out. For television sets with built-in teletext decoders, there is no reason why 24 rows of text cannot be displayed. The reduction or compression even improves legibility because the number of scan lines per inch is increased. A paper by Walter Ciciora of Zenith suggests that the only objection to raster compression is that when the text is superimposed over the normal video, the normal picture is also slightly compressed. Although this is barely noticeable, a solution would be also to compress the horizontal dimension or use a noncompression technique that would be required anyway for set-top decoders [4].

The noncompression techniques revolve around developing different ways to create characters (i.e., using fewer lines to do so). One way is to push lower case letters with descenders slightly above the normal line of text. This produces somewhat strange-looking lines of type, but the effect is often ignored after an initial feeling of surprise at the sometimes jumbled-looking display. The purpose, of course, is to eliminate the need for the extra line or two required to display the descenders. A better method retains descenders and uses only eight or nine lines per letter (depending upon the exact procedure) by employing alternate scan lines that are slightly different. This is known as character rounding and achieves the objective of a legible 24 rows (see Figure 5.2). Some forms of character rounding may require extra memory in the decoder, but other forms do not. A third technique that has been suggested is to use eight scan lines in one field and nine scan lines in the alternate field. This latter method would require twice the memory size of a decoder without character rounding, but it could create the best arrangement of dots and half dots (the rounding effect) for each character individually.

Despite these arguments for 24 rows, most of the teletext trials in the United States to date have elected to use equipment that displays 20 rows of text. CBS, Inc. has presented the FCC with a suggested teletext standard that incorporates a display format standard published by AT&T, and the latter standard defines the default page format as 20 rows of 40 characters [5]. Telidon Videotex Systems has also presented the FCC with a document that is the provisional Canadian teletext standard stating that the default page format is 20 rows of 40 characters; the Telidon filing similarly incorporates the AT&T standard [6]. (The CBS proposal also includes an optional 21st row of text, a ''service row,'' that can display up to 20 characters sent by the teletext system and 20 characters typed by the viewer at home.) The AT&T standard, on its own, was initially prepared as a display standard for videotex called a presentation-level protocol. The presentation level is one of the seven levels of the International Standards Organization's outline for a hierarchy of standards for interconnected systems. Because the AT&T standard was created to match the requirements of systems using a variety of display devices, character size can actually be specified within a wide range of sizes. In addition, there are specific commands to set three common page formats or character sizes (i.e., a display format of 80 characters on a row, a page format of 16 rows of 32 characters, and a page format of 20 rows of 40 characters) [7]. As will be described later in this chapter, the AT&T standard has since formed the basis of a videotex and teletext display standard to be adopted by the American National Standards Institute.

Figure 5.2 Example of character rounding.

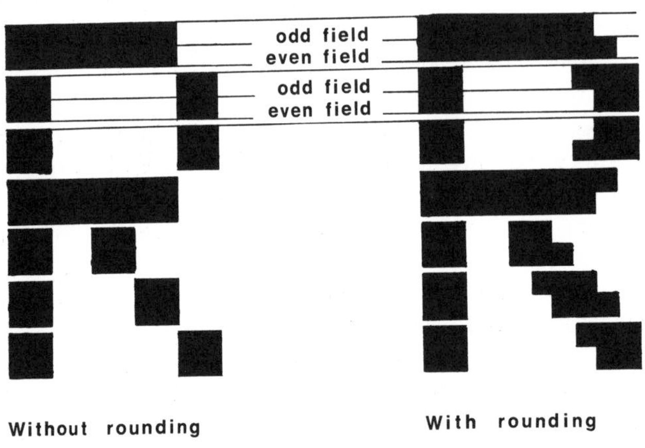

Serial/Parallel

Another controversy that has affected teletext and has a direct bearing on the appearance of the display is that revolving around the use of serial or parallel attributes. The attributes are the codes for color, size, texture, and the like for a character or row of characters. And the serial/parallel concern is not about how the characters are transmitted (which might be assumed by someone familiar with serial data transmission procedures) but rather about how the attributes are conceptually or physically stored in the decoder and the resulting effect on the display. The best way to describe what this means is to look at the history of teletext development. In the first British teletext system, the codes for size, color, and other attributes required as many bits as for a displayable character. For reasons of economy and efficiency of operation, the digital memory in the teletext-television set could hold only approximately the same number of characters as could be displayed on a single screen, in the same order. Therefore the attribute codes occupied a full character space in the memory. And as the memory was read to generate the display, blank spots would appear in place of the attribute codes. If a red word was to be followed by a yellow word, the attribute (the color code for yellow) would be in memory between the two words, but the screen would show the colored words with a normal space between them. This, of course, was no problem. The difficulty came primarily in creating graphic designs, where different colors often need to be side by side without an intervening blank spot. Consequently, the original teletext specification was soon changed to include a "hold graphics" code, which in effect would replace each attribute blank on the screen with a duplicate of the preceding graphic shape and color. The result was that many graphic designs began to look as they should, while still adhering to a serial attribute system. There are a small number of cases, however, where the hold graphics solution does not solve the problem; and the hold graphics code cannot be used for alphanumeric text. Thus in basic serial attribute systems, it is still not possible to put a red "a" next to a yellow "b" without one blank character space between.

Following the British experience with the serial attribute system, the French designed their teletext and videotex display format to solve the attribute blank problem by increasing the complexity of the decoder in the terminal or television set. The French solution was to decide that each character could be composed of more than the normal number of bits, with the extra bits defining the attributes for that character or graphic shape and the following characters on the row unless overridden by new attributes. This had the advantage of not only permitting different colored letters to appear side by side, but also allowing the system designers to attach an entire set of attributes to a character without any blank spaces appearing on the screen. On the other hand, additional memory is required in the decoder to store the attributes "in parallel" with the characters. (One or two attributes in the French system may cause a blank to appear on the screen.)

For several years, the British teletext (and videotex) system was the primary serial attribute system and the French teletext/videotex system was the primary parallel attribute system. And the two methods were incompatible. (Both the Canadian and the Japanese systems for teletext chose graphic methods unlike the mosaic graphics initially chosen by both the British and French systems, and thus the serial/parallel dilemma does not really apply to the Canadian and Japanese systems—see the following section on graphics.)

It is quite possible, however, to use the serial attribute system to create the parallel attribute effect, as has been demonstrated by British Telecom, the BBC, and the IBA for their respective videotex and teletext services [8]. Additional memory is used to store the attributes, and a cursor control code is used to "step back" into the "blank" space and write another character or graphic pattern [9]. This procedure is also compatible with the original serial attribute decoders to the extent that all alphanumeric characters and graphic patterns will appear properly, but some of the attributes may get overridden, and thus cancelled out, in the first-generation decoders.

Beyond that, it is also possible to create a teletext or videotex terminal that is functionally compatible with both the serial attribute and parallel attribute systems, as has been demonstrated by the European videotex standard ratified by the European Conference of Post and Telecommunications Administrations in May 1981 [10]. In fact, the European standard describes a reference model that can be implemented in several ways. One way is to use an eight-bit memory for each character, with seven bits required to define the character and the eighth bit used to indicate the presence or absence of attributes and other indicator bits stored in another eight-bit memory. Using the appropriate decoding procedure, a decoder could in almost all cases receive a page transmitted by either the British serial attribute system or the French parallel attribute system and display that page correctly.

For current teletext in the United States, the British-based systems are using serial attributes while the French-based systems are using parallel attributes. Often, given the high percentage of pages with normal text and simple graphics, there is no difference in the displays created by either procedure. The serial/parallel discussion has more or less fallen behind, or has been subsumed by, the discussion concerning the various ways to produce graphic designs on a television screen and the ways of arranging the data for broadcast transmission, both of which are explored below.

Graphic Formats

In the teletext world, at least four major procedures have been developed to create images or graphics on the television screen upon reception of a digital data stream. The simplest approach, in terms of design and the resulting display, is the mosaic approach. Any space the size of a normal character is divided into a number of smaller squares,

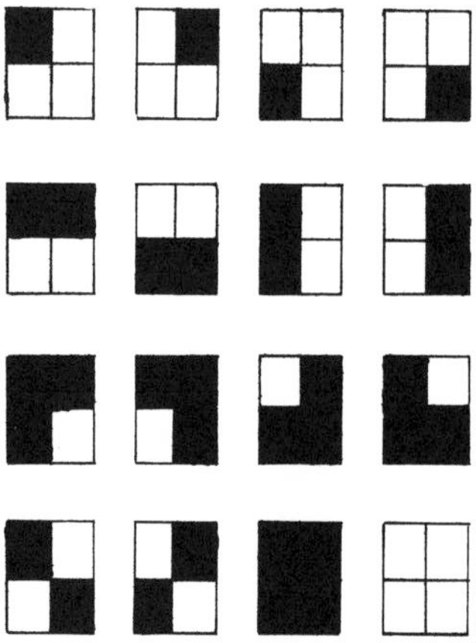

Figure 5.3. Mosaic patterns using a 2 × 2 graphic scheme.

or mosaics. The subsequent arrangement of these mosaic patterns creates the picture or graphic. A very simple mosaic method used by some display systems is to allow each character space to be divided into four smaller squares (see Figure 5.3). There are 16 possible patterns, and any design with curves or diagonal lines may end up looking rather crude. On the other hand, horizontal and vertical bars will appear as they should.

The advantage of the mosaic system is that each unique character-sized pattern is essentially just another character—a graphic character—and a rather small memory can hold the patterns as simply as the patterns for alphanumeric characters are held. At the lowest level, for example, only 52 patterns need to be held in the decoder's permanent memory to be able to display 26 capital letters, 10 digits, and 16 graphic patterns.

Mosaic-Six

The major teletext systems that employ mosaic graphics use a somewhat more flexible pattern of six possible squares per character space, sometimes called the mosaic-six standard (see Figure 5.4). Using six squares yields a total of 64 unique patterns. Moreover, these patterns can be displayed with adjoining squares touching each other (con-

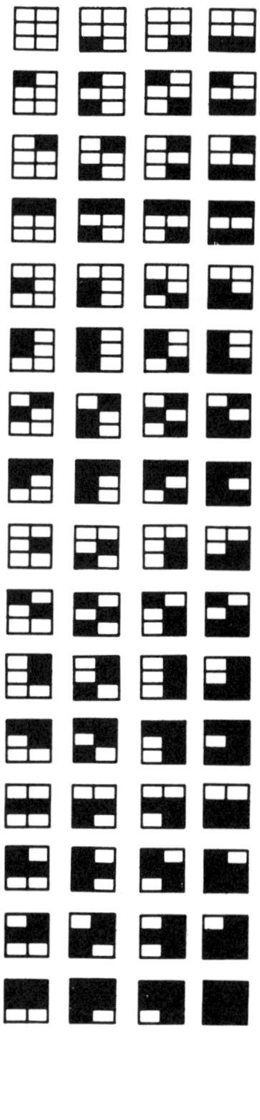

Figure 5.4. Mosaic patterns using a 2 × 3 graphic scheme.

tiguous graphics) or with a slight separation between each small square (separated graphics). The use of separated graphics can give curves a more rounded appearance than is possible with contiguous mosaic graphics, especially when viewed from a distance of a few feet or more.

The number of patterns in the mosaic-six approach is still within the economic limits of a character-generating chip, and in fact the addition of these graphic patterns to the character-generating capability of a decoder incurs a very minimal additional cost. The only complaint against mosaic-six graphics is that in some situations the display result is still too crude (i.e., a slanting line still looks like a series of little steps), and it is not currently possible to attach attributes to each of the little squares within a graphic character but only to the graphic character as a whole. Alternatives to mosaic-six graphics are considerably more expensive using current technology, and thus there is a substantial weight of opinion in favor of using mosaic-six graphics especially in teletext, based entirely on the projected relative low cost over the next several years. CBS, in its July 1981 filing with the FCC regarding teletext, indicated that simple graphics—mosaic-six—are a central part of CBS's planned teletext service, and that the more complex graphic techniques would be considered optional for teletext system operators [11]. All the 1981 broadcast teletext trials in the United States, with the exception of WETA–TV in Washington, D.C., used the mosaic-six method for creating graphic patterns.

DRCS

An alternative to the mosaic-six method is to use a mosaic with many more little squares per character space, although the total number of unique patterns begins to escalate rapidly. In reality, an upper limit on the number of mosaic squares is the number of dots in the character matrix used to create any character. One procedure that has been developed is to take the character space with, say, an 6 × 10 dot matrix and permit that character space to accept a new pattern of dots not stored in the decoder's permanent memory. In essence, the character space is being redefined, giving rise to the phrase "dynamically redefinable character sets," or DRCS. Thus the 2 × 3 mosaic becomes a 6 × 10 mosaic, or even a 10 × 12 mosaic, and the visual result is the ability to create fine-line graphics with proper curves.

However, DRCS has some present-day limitations that may affect the future viability of using the DRCS method to create graphics. For example, DRCS works best when only a small number of new patterns needs to be transmitted to the terminal or television set, and these new patterns would then be used repeatedly in the displays during a given session at the terminal. Otherwise, a great deal of time is required to transmit new patterns bit by bit (a 6 × 10 character matrix, for example, would require at least 60 bits to define a new pattern with only two possible colors in that pattern). In short, DRCS works best with "downloading" a set of nonstandard characters, such as a graphic pattern or an alphabet for a little-used language that is to be used for an entire group of pages to be immediately viewed.

There are still limits on the number of patterns that can reasonably be stored on a permanent basis in a decoder and on the time that can be devoted to transmitting new DRCS patterns without exhausting the patience of the viewer. The British proposal for DRCS, for example, suggests that only 94 DRCS characters be available per page (i.e., only 94 DRCS patterns can be stored at a time) and that the same 94 patterns be used throughout a series of pages. In the teletext mode, the DRCS patterns would be transmitted to a 94-character memory in the decoder prior to transmission of the pages using these DRCS patterns. The CBS, Telidon, and AT&T proposals for DRCS suggest a similar limit (96 characters) on the number of patterns to be stored in the receiver at any one time.

In addition to these limits of DRCS as a means of creating fine-line graphics, additional limitations may be imposed by the way DRCS is defined in the AT&T videotex standard adopted by a number of organizations in the United States for teletext. A paper by John Lopinto of Time, Inc., points out that the AT&T protocol states that DRCS characters may vary in size up to the size of the screen, thus possibly greatly expanding the number of bits required to define each DRCS pattern; moreover, in the AT&T specification, the DRCS pattern can be defined only if the decoder can also interpret geometric drawing instructions (which are discussed in the next section on geometric graphics) [12]. This not only requires a more complex decoder but also increases the number of bits required to define a pattern. As Lopinto points out, in teletext an increase in the number of bytes per page slows down the broadcast cycle, and either the size of the teletext magazine must be reduced or viewers must accept a longer maximum cycle time.

A successful use of DRCS, however, may be an application where graphic patterns are transmitted to an intelligent terminal such as a microcomputer for subsequent use as part of a program or session. For example, Playcable uses the DRCS technique to send the graphic patterns needed for a given game to the Mattel Intellivision prior to the start of each game. Because no broadcast teletext systems are actively using DRCS, though, the broadcast applications of DRCS remain to be tested in the marketplace.

Geometric

The mosaic graphic techniques are essentially an outgrowth of character-generation techniques, because mosaic patterns are treated exactly as alphanumeric characters. Similarly, another graphic technique in teletext is essentially a variation on earlier procedures used in a related area (i.e., the so-called alphageometric graphic in teletext is the offspring of previous computer graphics procedures). A standard text on computer graphics explains that in the simplest computer graphic systems, patterns can be created using a small set of primitive graphic functions [13]. In other words, five simple functions—for moving the terminal's electron beam (which is aimed at the screen) a given distance or to a point, for drawing a line a given distance or to a point, and for displaying a point—can be utilized to draw any line image on a CRT screen.

In the mid-1970s, a group of researchers working within the Department of Communications in Canada proposed a videotex system that would use a set of graphic functions, called Picture Description Instructions (PDI), to create graphic designs rather than using mosaics [14]. Instead of a set of mosaic patterns being permanantly stored in the decoder's memory, a microprocessor in the decoder would receive drawing instructions and create the image as instructed. The Canadian system specified seven PDIs, plus a TEXT instruction as listed in Table 5.2. The POINT command, for example, would contain within it a screen location and would cause the terminal to display a point at the given location. The ARC command would contain within it the end points of the arc and its radius, and the displayed arc could optionally be drawn as a simple arc, an arc with end points joined by a straight line, or a solid arc. The LINE command would create a line based on its end points, and so on (see Figure 5.5). In this way, a small set of PDIs could be used repeatedly to define pictures, from the simple to the complex. In addition to the drawing instructions, the BIT PDI specified that the subse-

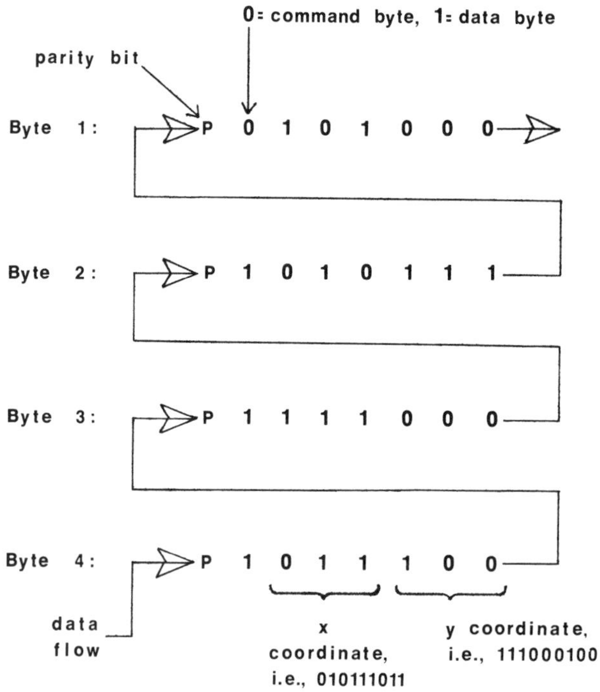

Figure 5.5. Example of the construction of a PDI code.

quent string of bits would define a pattern dot by dot, such that every four bits might define the color of each of the smallest possible dots visible on the television screen. In this way, a photographic-quality picture could be transmitted. The seventh PDI, CONTROL, was used to transmit the attributes of a pattern (e.g., setting the color of an object).

As the Canadian Telidon system was developed, it was obvious that the use of PDIs, or the alphageometric approach, permitted graphic images to be displayed that were much finer than those possible with mosaic-six graphics. A secondary advantage was that a simple pattern could be defined using only a few PDIs, while the same pattern in a mosaic-six system would require an entire mosaic of graphic characters. Additionally, it was argued, images transmitted as PDIs could be displayed with as fine a resolution as each new generation of television sets or terminals could manage, while a mosaic-six pattern would always be a mosaic-six pattern regardless of the resolution capability of the terminal. On the other hand, a Telidon receiver needs a microprocessor and additional memory, and a complex image composed of PDIs may actually require three or four times the number of bytes needed to represent the same image using mosaic-six graphics.

In 1981, AT&T published its version of a videotex display standard that incorporated both mosaic-six graphics and PDIs. Although based heavily on the Canadian PDI structure, the AT&T arrangement was slightly different (see Table 5.2). As a whole, the AT&T standard describes two sets of digital codes: graphic codes and control codes. The graphic codes are then divided into two subsets: text and PDIs. The "text" set includes not only alphanumeric characters but also the mosaic-six patterns and DRCS

Table 5.2. Picture Description Instructions

Early Canadian PDIs[a]	AT&T PDIs[b]	
Point	Point	
Line	Line	
Arc	Arc	graphic
Polygon	Polygon	primitives
Area	Rectangle	
Bit	Incremental	
Text	Reset	
Control	Domain	
value	Text	
status	Texture	
domain	Set color	
blink	Select color	
transparent	Control status (wait)	
line texture	Blink	
fill		
wait		

[a]*From:* H. G. Bown, C. D. O'Brien, W. Sawchuk, and J. R. Storey, A Canadian Proposal for Videotex Systems: General Description, Communications Research Centre, Ottawa, Canada, November 1978.
[b]The first six PDIs, the graphic primitives, can have four forms, so there are actually 32 PDIs in the AT&T standard.

capability. The PDI set contains the geometric primitives and the control commands listed in Table 5.2. At the same time that this was happening, the British and French also included DRCS and alphageometrics as enhancements to their original systems.

Concurrent with the publication of the AT&T specification or soon thereafter, a number of major companies experimenting with teletext and videotex in the United States announced that they would support the AT&T standard as an industrywide standard. In addition, the Telidon system itself was adapted to conform to the AT&T specification, and both Canadian and French teletext advocates (i.e., Telidon Videotex Systems and CBS) petitioned the FCC to adopt the AT&T standard for the display of broadcast teletext.

Finally, during 1982, the standards situation in the United States began to sort itself out somewhat owing largely to these actions and to the activities of four other groups: the American National Standards Institute (ANSI), the Canadian Standards Association (CSA), the Electronic Industries Association (EIA), and a group called the Videotex Technical Experts Panel. In early 1982, an ANSI subcommittee known as X3L2.1, composed of about 60 members from a variety of U.S. companies, began working with a similar CSA subcommittee to arrive at a North American standard for both videotex and teletext that would be acceptable as a voluntary standard in both the United States and Canada. The ANSI group took as their starting point the AT&T specification and began resolving minor ambiguities within the document and discrepancies between that document and the Canadian Telidon specifications.

While that was going on, a subcommittee of the EIA was also trying to define a subset of the AT&T specification for use in broadcast teletext. (It was readily agreed that a practical and economic broadcast teletext decoder would not be able to implement the full AT&T specification.) After discussions between the X3L2.1 group and the EIA "BTS Teletext Steering Committee Special Working Group," it was decided that the results of each group's effort would be combined into a single set of standards. Therefore the ANSI standard, now called the "Videotex/Teletext Presentation Level Protocol Syntax," or simply the North American PLPS, contains as an appendix a General Service Reference Model for teletext supplied by the EIA committee. During 1982, the ANSI videotex/teletext standard began being processed through the various levels of ANSI prior to being declared an ANSI standard. A similar process was begun within the Canadian Standards Association.

For international purposes (i.e., beyond North America) the ANSI/CSA standard was submitted to the Videotex Technical Experts Panel, a group that had been established under the auspices of the U.S. State Department for the purpose of providing advice to the State Department regarding a U.S. position on videotex standards. The State Department is to present the North American PLPS in late 1982 to the CCITT, an international telecommunications standards organization, for consideration as a world standard (either alone or in conjunction with other videotex standards accepted by the CCITT).

Because the North American PLPS and the teletext subset contain both mosaics and PDIs, either method could be used by teletext operators adhering to the standard, depending largely on matters of cost and availability of equipment. For broadcast teletext, it is still debatable whether mosaic patterns are more or less economical than alphageometric graphics for several reasons for the near future. Only one television station in the United States, WETA in Washington, D.C., is testing a teletext system in which all graphics are created using PDIs, and a number of disadvantages have appeared [15]. First, the use of PDIs can boost the average number of bytes per page to a level higher than the average for pages using mosaic graphics, at least under current practices. In fact, while a page with mosaic graphics simply cannot contain more than 800 bytes (a 20×40 format), a page with PDI graphics may contain 2,000 to 3,000 bytes. Unless the graphic artist knows all the implications of using each PDI, the artist may unwittingly increase the number of bytes per page, for example, by building a grid by repeatedly using the "line" command rather than the "texture" and "rectangle" commands. Secondly, preliminary research seems to indicate that the use of simple color and graphics substantially improves a teletext display over a display of black and white text, but that the addition of more complex graphic patterns and colors does little more, except to cause both the graphic artist and the viewer to spend more time to create and receive the page, respectively [16]. On the other hand, judicious use of alphageometric techniques can result in pages with many fewer bytes per page than in a strict mosaic graphics system.

Within the cable environment, however, the alphageometric approach may be viable much sooner, because the digital information can be transmitted using full-field teletext (effectively reducing any additional time required to transmit the more complex graphics), and the cost of the decoder can be met in a number of ways. The teletext decoder can be added to the set-top cable converter already used by many cable systems, or a teletext decoder located within the cable system but outside the home could serve more than one user, or a single teletext decoder at a cable headend could be used to display rolling pages of teletext on a dedicated channel.

Photographic

At the high end of the graphic scale in teletext is the ability to digitally convey full-color, photographic-quality pictures. The British, French, and Canadian teletext/videotex systems all have demonstrated the alphaphotographic technique, and the AT&T/

ANSI specification contains an INCR.POINT command that allows an image to be defined as a series of color dots that will achieve the alphaphotographic effect.

Actually, several methods are used to digitally create teletext pictures classified as alphaphotographic, or simply photographic. The commands such as BIT and INCR.POINT within the geometric schemes can, for example, achieve the photographic effect, although not necessarily in full color. A slightly different procedure is used in Japan, where the individual dots, or picture elements, can be in color but only in one of eight colors. Another approach used by the British is to determine the luminance and chrominance values for each picture element and then to create a compressed digital data stream by transmitting only the difference between the actual value and the estimated value based on previous values (differential pulse code modulation) [17].

The photographic techniques, however, require substantially more bytes per image than the geometric techniques, and again, in broadcast vertical blanking interval teletext, the viewer either waits longer to see the page or the broadcast cycle must be reduced. Additionally, the alphaphotographic decoder is much more complex than a mosaic-six decoder, and British engineers estimate that an alphaphotographic decoder may require at least 25 times the memory of a mosaic-six decoder [18]. Typically, in the experimental uses of alphaphotographic teletext, only a portion of the screen is devoted to the photographic image in order to minimize the amount of digital data that needs to be transmitted or broadcast. The remainder of the screen is used for text or mosaic graphics.

Given the various approaches to graphics in teletext, and despite the historical development of one or another technique in different countries, it seems that all major systems incorporate all the forms of graphics. The Canadian Telidon system's PDIs, for example, can be interpreted by a decoder that creates mosaic-six patterns from PDIs, while the British teletext standard contains all the forms of graphics in a hierarchical structure, and the North American PLPS contains all the forms of graphics in a unified structure revolving around the use of commands that are conceptually PDIs themselves. The development of all forms, or the dominance of one or another, depends upon the desire or ability of the semiconductor companies and the teletext–television manufacturers to produce the chips and decoders, and the teletext operators to produce services, such that consumers are willing to pay the price of the equipment.

Transmission Format

Parallel to the controversies over page formats and graphic capabilities has been considerable discussion of the most appropriate procedure for placing the digital information within the television signal. Questions of how to shape the data pulse and how many data pulses to place within a single scan line have yet to be definitively answered for NTSC television, but the primary debate has been over the relative merits of the so-called fixed format system developed in England and the variable format system developed in France for teletext transmission. Ultimately, there is probably no effect that cannot be accomplished using either method. The primary arguments that the British advance in favor of their fixed format are that the resulting decoder design is less costly to implement and the broadcast data are less susceptible to interference without the overhead of complex digital error correcting codes. The arguments for the variable format are that it is more flexible and more complete, with provision for a wider range of applications.

The fixed format concept means essentially that there is a direct one-to-one relationship between the position (or timing) of a character or digital code within the television scan line and the position of that character or code in the teletext–television set's digital memory and on the set's screen [19]. There are two principal benefits of utilizing this technique. First, this positioning requirement means that the digital signal is locked to the most stable part of the television signal—the horizontal synchronization pulse. Second, no microprocessor is needed in the decoder.

In contrast, the variable format teletext systems have no such relationship between the location of the data in the television signal and the resulting display. Consequently, a microprocessor is needed in the decoder to interpret the incoming digital stream and assemble the appropriate display. However, the presence of the microprocessor means that additional error-correcting procedures can be introduced and accommodated within the teletext receiver.

One of the reasons that the U.K. fixed format system seemed initially unsuited for the United States was that there was an obvious connection with the broader bandwidth of the European PAL standard television signal (i.e., a single scan line can hold 40 characters—the same number displayed per row—plus a few bytes for row and line numbers). Specifically, the U.K. teletext standard calls for the data content of a scan line to be 45 bytes of eight bits each [20]. Forty bytes are used for the 40 characters per row, three bytes are used at the beginning of the line to ensure that the timing of the data is received accurately (two bytes for clock run-in, and one byte for the framing code), one byte contains the magazine number, and one byte contains the row number. The magazine and row bytes are protected by a Hamming code, which means that only half the bits in the string are real data, and the other half are used by the decoder to detect errors in the data bits. (Appendix A contains a Hamming code table as part of the description of the Canadian teletext transmission standard.)

The television signal in the United States, narrower than the television signal in Europe, cannot carry the same number of bytes in the scan line using the same technique. Nevertheless, it is still possible to use the fixed format with a display of 40 characters on a line by using a procedure called "gearing" [21]. In simple terms, the character data for each row are divided into subgroups and transmitted on separate scan lines. This subdivision can be done in a number of ways. For example, the number of character bytes per line can be set at 30, and the first three rows are transmitted with 30 characters each, and then a line of 30 characters is transmitted that in reality contains the last 10 characters of each of the first three rows. The 30/10 gearing is in fact used by Southern Satellite Systems to transmit their CableText service within the WTBS television signal, and this yields a burst data rate of 5.5 megabits per second. Another possibility is to transmit four lines of 32 characters, and then a line of 32 characters containing the last eight characters of each of the preceding four rows. This 32/8 gearing is used by WFLD–TV in Chicago and yields a burst data rate of 5.727272 megabits per second.

For use in the United States, the U.K. Teletext Industry Group has proposed that their fixed format system operate at a burst data rate of 5.727272 megabits per second with a total of 37 bytes per line, including the five introductory bytes for clock run-in, framing, magazine number, and row number. The fixed format system, both in England and in the United States, can also achieve the effect of parallel attributes, PDI graphics, and the like through the use of nondisplayable rows, which have been called "ghost" rows.

The variable format system used by NBC and CBS and proposed by CBS and Telidon Videotex Systems to the FCC in 1981 also operates at 5.727272 megabits per second, although the maximum number of bytes per line is 36 rather than 37, and eight of the bytes are used for establishing the timing and for carrying additional information about the character data or digital codes on the line. (Appendix A contains the provisional Canadian broadcast teletext specification, which is identical to the Telidon/CBS system proposed for the United States.)

Examining the structure of the transmission format will give a good overall view of how digital information is actually placed within a television signal. Each scan line is preceded by a line synchronization pulse followed in color television by a color burst, followed by a modulated signal. To impress data onto this signal, the modulation is restricted to two possible levels: the black reference level and a level nearer the white reference level (see Figure 5.6). In most of the systems proposed, the latter level is approximately seven-tenths of the white reference level. This higher level indicates the presence of a binary "1"; when the signal is at the lower level a binary "0" is indicated. The particular coding technique used by virtually all teletext systems is Non-Return-to-Zero (NRZ) coding (i.e., if the digital stream contains a series of ones, the modulated signal remains at the higher level because in all systems the duration of each "1" is fixed and the continuous signal can thus be interpreted correctly).

The clock run-in, or clock synchronization bits, is also a common feature in all teletext systems, because all systems require accurate timing so that the bits can be identified as such. In the CBS system, for example, the clock synchronization bits are the first 16 bits, which appear in the scan line beginning approximately 10.5 microseconds after the beginning of the horizontal synchronization pulse. These 16 bits are alternating ones and zeros, as depicted in Figure 5.6, and enable the decoder to lock onto the timing and amplitude of the digital data stream.

The next eight bits (the third byte) represent the framing code and allows the decoder to know when a byte begins. A number of possible codes (i.e., arrangements of ones and zeros) can be used to accomplish this, and the CBS and Telidon proposals specify "1 1 1 0 0 1 1 1," with alternate codes reserved for future use. The U.K. proposal for

Figure 5.6. Impressing digital information onto a television scan line.

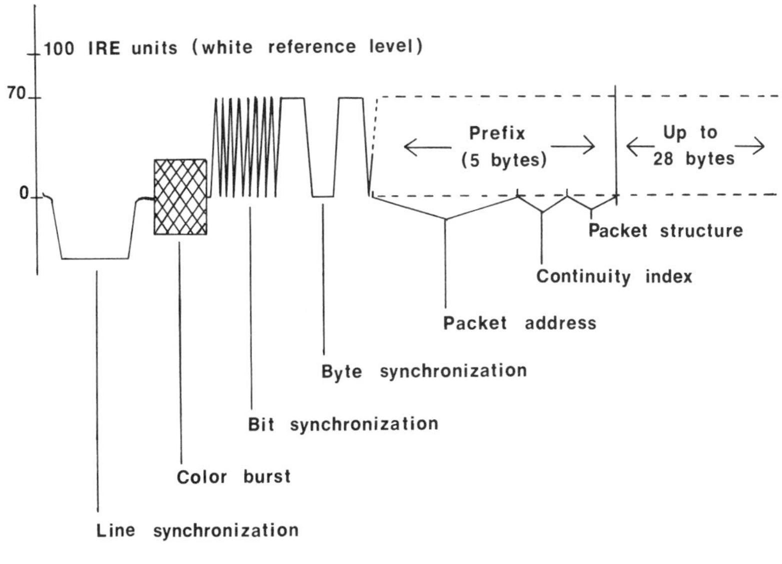

U.S. teletext suggests "1 1 1 0 0 1 0 0" as the framing code. The choice of the framing code is based upon the need to find a code that could be correctly interpreted by the decoder even if one of the bits is received incorrectly.

The subsequent five bytes in the CBS and Telidon North American Standard are called the prefix, composed of three packet address bytes, a continuity byte, and a packet structure byte. The three address bytes are Hamming-protected; thus only 12 bits are real data and the other 12 bits are used to detect errors according to the Hamming algorithm. This gives a total of 4,096 possible addresses, which correspond to the magazine and row numbers in the British system (which has a maximum of 256 possible addresses). These addresses have also been called "channels," to convey the idea that 4,096 time-divided data channels are available within the teletext system, for whatever reason the digital codes might be used.

The fourth byte in the prefix, the continuity index, is used to detect the loss of a line of code owing to interference. This byte, also Hamming-protected, allows consecutive lines of code with the same address to be numbered sequentially from zero to 15. If the decoder notes a gap in the sequence, one or more lines of code, or packets, have been lost.

The final byte of the prefix indicates whether the scan line contains the maximum number of bytes, whether suffix bytes are included, and whether this is the beginning of a data group. A data group consists of all the lines with the same address; at the start of a data group, the first eight bytes specify such things as the number of data blocks (i.e., lines of data) within the group, the size of the last block in the group, a routing byte (e.g., for network transmission purposes), and continuity and repetition codes (i.e., further checks on whether all blocks in the group are received).

The prefix is followed by up to 28 bytes of character data or other digital codes. The final bytes—as many as the last three—may optionally be used for further error-detecting or error-correcting purposes. In other words, if these suffix bytes are used, the actual number of character bytes per scan line is 25.

The character bytes in virtually all teletext systems are seven-bit codes with the eighth bit as a parity check, a standard error-detecting procedure for data transmission. These character bytes can be any valid code within the standard being used, whether the codes are to be interpreted as alphabetic characters, mosaic patterns, PDIs, and so on. (Appendix C contains the defined reference tables for the 128 unique codes that are possible using a seven-bit code, as contained in the ANSI standard. The ANSI standard also includes an eight-bit environment, which gives 256 unique codes that can similarly be interpreted in a number of ways depending upon the reference table being used.)

In the CBS teletext system (based on the French Antiope/Didon system), a data group corresponds to a page or "record." Each record is preceded by a record header that may contain the following codes:

record type (e.g., cyclic broadcast teletext or a targeted message)
record link (i.e., to other records)
record address
address extension
classification codes
record header extension.

The address and extension codes are Hamming-protected to decrease the chance of faulty reception. These bytes that form the record header are part of the character data that follows the prefix.

Although in the CBS system for broadcast teletext there is a one-to-one relationship between a data group and a record, there is a provision, called "segmentation," for identifying sections of the teletext page (which may, in fact, contain more than one record). Using the segmentation scheme, the sections of the page are called Session Service Data Units, and they can be bounded by using a number of the codes for positioning the cursor or for selecting certain attributes. One or more of the Session Service Data Units can also be identified as a Quarantine Unit, which means that all the data in the Quarantine Unit must be accurately received before processing or display. The purpose of the segmentation is to increase the capability for protecting teletext pages from the adverse effects of spurious errors introduced during the transmission process.

Having outlined the U.K. teletext system and the CBS-sponsored system, we find it worthwhile to consider two other, different, systems that could affect teletext development in the United States.

The first of these is the line 21 system, which is a variation of teletext used in the United States for closed captioning and for scrolling text services, as described in previous chapters. In fact, the line 21 system is (as this is being written) by far the most prevalent form of teletext in use in the United States, and the system providing the most closed captioning anywhere in the world. Other countries that have a greater number of teletext–television sets in use do not as yet engage in as much closed captioning activity.

The most obvious difference between the line 21 system and the other teletext systems, aside from the lack of selectivity on the part of the user, is the very slow data rate [22]. In the line 21 system, only two characters are transmitted per scan line, and only one field is used (see Figure 5.7). Therefore, instead of 28 or 32 characters per scan line, less than one-tenth that number is being transmitted. Consequently, the pulses forming each bit are larger in a sense, although the amplitude of the pulses is less than that proposed in the U.K., CBS, and Telidon systems (i.e., approximately 50 instead of 70 IRE units). The synchronization and framing functions are accomplished with a seven-cycle 0.503 MHz pulse followed by two "0" pulses and a "1" pulse. The last

Figure 5.7. Data signal format for the line 21 system.

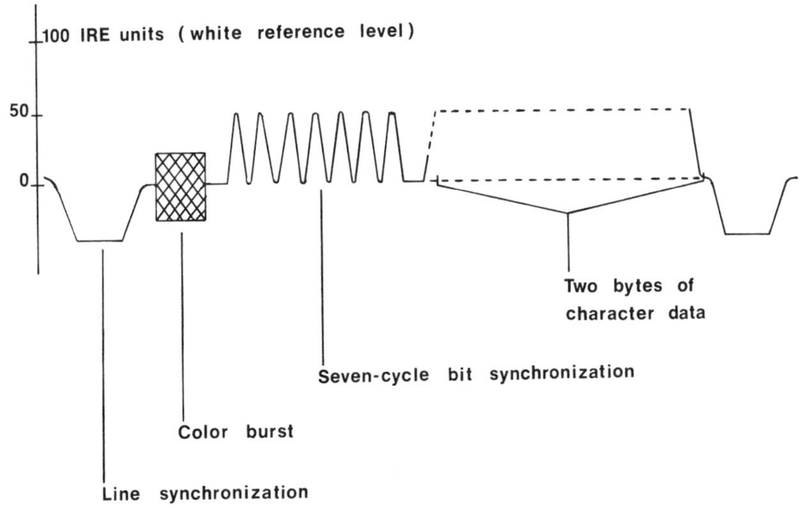

two cycles of the synchronization or clock run-in are taken together with the three digital pulses to form the framing code that precedes the actual character data.

The line 21 system uses a set of control codes to begin and end captions and to select attributes, such as color. (Appendix B contains a complete list of the control codes.) The control codes are always composed of a two-character sequence and are always transmitted twice. Therefore, because only two characters are transmitted per scan line, each instance of control information occupies two transmitted frames of video (i.e., two occurrences of line 21, field one). The control codes that begin a caption or line of text are used to designate colors, underlining, indentation, italics, and row numbers. Within a given caption, control codes can be used to change color, underlining, or italics. The end-of-caption code in normal operation causes the decoder to switch back and forth between two digital memories, one holding the current caption being displayed and the other receiving the new caption coming in.

The text transmitted by the line 21 system can appear on the screen in several different ways. In normal caption operation, captions appear either on the top or bottom four rows of a screen that can hold 15 rows of text if full. The captions appear cut into the normal video, set within a black box. In the caption roll-up mode, the text may scroll within the bottom four rows (i.e., new characters are always written on the 15th row, and preceding rows of text are bumped up one, to be erased at row 12). In the "video-text" mode, all 15 rows of text are used. Characters first appear at the top, in text row 1, but as 15 rows of text appear, the lines of text begin to scroll upward. The scrolling effect can be altered, if desired, by using a "restart" command that erases the screen and starts the text at the top again.

Because the line 21 system was developed in full knowledge of the teletext development work going on in England and later elsewhere, the immediate question is why was the line 21 system designed to be so much unlike the more complex teletext systems. The answer lies in considerations of cost and design simplicity. At the time the line 21 system was being implemented, the chief concern was to produce a closed-caption system that would be inexpensive enough for large numbers of hearing-impaired people to afford, and easy to use. It became apparent that a decoder without a microprocessor would be more economical, and that the relatively slow data rate could be handled by available technology. Furthermore, there was no need to incur the expense of adding selectivity of pages because the system was primarily intended for closed captioning.

The resulting line 21 standard, authorized by the FCC in 1976, affects the display and use of the system in several ways. For example, while color codes are used, color is displayable only if the decoding circuitry is built into the television set; if a set-top decoder is used, no color appears (i.e., the characters are white). Actually, given the limits of the line 21 system, a set-top decoder could display the horizontal portion of a character in color but not the vertical portion, so the color capability was eliminated from the set-top decoder. A second consequence of the line 21 system is that television programs containing captioning data, or any such data on line 21, can be relatively easily recorded on videotape. The ability to record those data on videotape is essential to the captioning process, where programs must be viewed again and again as the intellectual effort and the physical effort in creating and reviewing captions takes place; videotapes containing the captioning data in line 21 can then be easily sent to the broadcast networks or individual television stations. Attempts in the United States to videotape programs with high-speed teletext data in lines 15 and 16 have met with only limited success.

The status of the line 21 system is in doubt because of the FCC's pending decision to permit any kind of teletext on all the available lines of the vertical blanking interval. The closed captions or text on line 21 would not be protected nor would the line 21 system be the only teletext system available for encoding and transmitting captions. As this is being written, line 21 television sets and line 21 set-top decoders are the only teletext (or pseudoteletext) receivers commercially available. The receivers are sold through Sears, Roebuck and Company, and nearly 50,000 have been purchased. No one is quite sure whether the line 21 system will continue to exist alongside the more complex teletext systems or whether it will gradually cease to exist as closed captioning is accomplished through the other forms of teletext.

Another teletext system worthy of mention is the Japanese system, if only because of the growing influence of Japanese manufacturers in the computer and electronics industries and because the Japanese television standard is the same as that in Canada and the United States.

The Japanese system is similar to the U.K. and CBS systems proposed for North America on a very general level, but on the detailed level there are several differences in the transmission format, and major differences in the display procedure [23]. As is the case with the U.K. and CBS systems proposed for the United States, the Japanese use a burst data rate of 5.727272 megabits per second. In fact, it was apparently only after the Japanese investigated the 5.72-megabit data rate that systems in the United States began to try that same data rate [24]. The Japanese teletext system also utilizes three initial bytes for synchronization and timing, but here slight differences begin to appear. The Japanese specification calls for the data to begin 9.78 microseconds after the start of the horizontal synchronization pulse, while the CBS system specifies 10.5 microseconds (in each case there is a tolerance of about one-third of a microsecond). And while the first two bytes, for bit synchronization, use the pattern of alternating ones and zeros in both cases, the third byte has a "1 1 1 0 0 1 0 1" pattern in the Japanese system and a "1 1 1 0 0 1 1 1" pattern in the CBS system. The Japanese system then uses three bytes for packet identification (compared with five bytes in the CBS system) followed by a packet or scan line of up to 31 bytes (compared with 28 bytes in the CBS system). Page address and display addresses are part of the data content of certain types of packets in the Japanese system. Of the three bytes for packet identification, one is a service or system identification and interrupt, and the other two are for display features and packet identification. Essentially there are seven types of packets, as listed in Table 5.3. One of the packet types allows horizontal scrolling of text on the screen (vertical scrolling is also possible).

The Japanese display format divides a page or screen into 248 × 204 dots, which can be referenced either as dots or as grouped blocks of color so that there are 17 rows of 31 color blocks each per screen. A complete data packet—one scan line of data—is required to transmit the dot pattern for one line of dots on the screen. However, in the color block mode, one scan line can carry the data for one row of blocks, which is actually composed of 12 lines of dots. The standard formats for characters are either 8 rows of 15 characters or 16 rows of 31 characters.

Unlike all the teletext systems tested thus far in the United States, the Japanese system currently does not use a character-generating chip in the decoder to produce the dot patterns related to the seven-bit ASCII code for characters. Instead, the dot patterns themselves are transmitted, and a memory of approximately 5 kilobits is required in the decoder to hold a screenful of dot patterns. In contrast, this is about nine times the memory required for the current generation of U.K. teletext decoders.

Table 5.3. Data Structure in the Japanese Teletext System

Packet type[a]	Function
Page control	Page addresses and displays control codes
Color code (A)	Set the color of letters and graphics
Color code (B)	Set background color
Pattern data	Contains pattern data
Horizontal scroll data	Contains pattern data (including text) to be horizontally scrolled
Program index	Transmits program numbers
Dummy	Contains no information data

Based on: NHK Technical Research Laboratories, Japanese Teletext, March 1981.

[a]A packet is the amount of digital data transmitted on a single scan line, containing up to 34 bytes.

In sum, there are numerous ways of creating formats for transmission and display of teletext, and each may have desirable features. Standards, when established, may overlook the benefits of one or another system, and as technology changes, new features are possible, so that eventually new standards may be needed to continue to produce better systems.

Standardization

The standardization dilemma is that some general agreement among manufacturers is necessary to permit economies of scale, but that once established, the existence of the agreements or standards may inhibit optimum development of the technology. This is particularly true in systems relying on semiconductor technology, on the manufacture of chips of densely packed integrated circuits. The design and production costs are so high that only when the chips are in high-volume production does the entire process become economical. Teletext systems are based on computer chips to store digital data and to generate displays, among other functions.

It is easy to say that teletext standards should be specific enough to allow different manufacturers to produce compatible systems but general enough to allow different systems with different features to compete in the marketplace so that the competition will encourage enhancements. In reality, however, it has been very difficult to know where to begin—and end—teletext standardization. A statement by an FCC commissioner has suggested that the data rate, for example, is a technical standard that can be set without hurting teletext development [25]. And the data rate of 5.72 megabits per second seems to be the preferred choice. On the other hand, the much slower data rate of the line 21 system has advantages, and a preliminary report from the WETA–TV teletext trial, where a data rate of 4.58 megabits per second is used, concludes that the higher data rates should be studied further and that 5.72 megabits per second specifically may be too high [26].

Even the simple matter of deciding which lines in the vertical interval are available for teletext transmissions is not without controversy, because the line 21 system gained an early foothold on line 21 but then faced challenges from other systems seeking to use that space for higher-volume teletext.

Aside from the technical issues, there are suggestions that some sort of standardization might be beneficial regarding the appearance of a teletext page and the consumer's interaction with teletext. For example, the name of the publisher of each teletext page could be required to appear on each page, and even perhaps to appear in a certain part of each page.

Internationally, there seems to be a growing division between Western European and North American standards. It seems unlikely that there will be a single world standard for teletext at any time in the near future. To a large extent, this stems from existing differences in technical systems for television itself. And while it may be possible to hammer together a "standard" that would incorporate the technical differences of all television–teletext systems, there is no compelling reason to do so. On the international level, individual markets may be large enough to support the development of teletext systems unlike systems on the other side of the world. Furthermore, international differences may be the catalyst for improvement; already the systems proposed for the United States seem to have benefited from the competition among other countries [27].

One of the key considerations in setting technical standards is to permit systems to develop that will not render older pieces of equipment obsolete in a few years. The prime example is color television, where a black and white receiver can correctly display a color television signal (minus the color), while a color receiver can correctly display both a black and white and a color signal. In teletext, if it is true that the first system to be widely successful will incorporate less complex (and therefore less expensive) decoders or teletext–television sets, then any technical standard should permit the older teletext-television sets to be still useful even when new features are added that are visible only on more complex receivers. Both the U.K. filing and the CBS and Telidon filings before the FCC have argued that their respective systems permit this sort of growth.

Other technical issues that could seriously affect the success of teletext have hardly been mentioned in the arguments over standards. There is, for example, a problem with ghosting, or multipath reception of television signals. In geographical areas where a television picture is still viewable, albeit with "ghosts," the multiplexed teletext data may be considerably disrupted—and therefore unreadable—because of the same multipath reception. As one Electronic Industries Association paper points out, television reception in the United States is generally inferior to that in Europe, and teletext systems technically successful there may not be so successful here [28]. Consequently, to solve the disruption caused by multipath reception, U.S. teletext–television sets may require a "time domain equalizer" that, it is noted, is under development in Japan and in Europe but not yet in the United States.

Adoption of some sort of broadcast teletext standard is generally believed to be a sine qua non for teletext development. If a standard is not promulgated by law, or widely accepted voluntarily, it seems likely that teletext will develop only within certain areas (i.e., for applications where users are willing to pay the cost of the relatively expensive teletext receiver). If so, then more than one standard could exist (e.g., one for closed captioning, one for stock market or business information systems, one for a national weather service, and so on). In the cable television environment, where different standards may easily exist, a technical standard could eventually evolve to compete with the standard for broadcast teletext because of the low cost associated with a high-volume of semiconductor chips.

Although it is still too early to tell for sure, it seems likely that for broadcast teletext the Teletext Service Reference Model of the ANSI North American PLPS will be accepted as a voluntary standard. Bernard Lechner of RCA, chairman of the EIA committee investigating teletext standards for about three years, has said that the FCC's approval of a ''marketplace'' approach was really based on an expectation that an industry group, and not an abstract marketplace of consumers, would agree upon a standard [29]. Yet it is also necessary to point out that while the EIA may agree, the standard is far from unanimous and will always be only voluntary unless the FCC steps back in.

References

1. Lipoff, Stuart, Lopinto, John, and Siedel, Robert, Final Report of the Ad Hoc Page Format Working Group, Task Group A—Systems, Teletext Subcommittee, Broadcast Standards Committee, Electronic Industries Association, August 24, 1981. For a good description of the horizontal resolution limits and the effects of color in the NTSC television system, see also W. C. Treurniet, Display of Text on Television, CRC Technical Note no. 705–E, Ottawa, Canada; Department of Communications, 1981.
2. Crowther, G. O., Teletext and Viewdata Systems and Their Possible Extension to Europe and the USA, *IEEE Transactions on Consumer Electronics* CE–25(3): 292 (July 1979).
3. Ciciora, Walter S., Twenty-Four Rows of Videotex in 525 Scan Lines, unpublished paper, Zenith Radio Corporation, 1981; Carl G. Eilers, Recommendation for Standards of Character Dot Matrix Consistent with DRCS, Page Format for System M, and Typically Used Color CRTs, *BTS/Teletext Subcommittee Interim Report,* vol. II, Electronic Industries Association, Washington, D.C., 1981, pp. 268–279.
4. Ciciora, W. S., Twenty-Four Rows in 525 Lines, the Mini Gateway, the Micro Gateway, and the Real World, *Videotex 81,* Online Conferences Ltd., Northwood, England, 1981, p. 561.
5. See CBS Television Network, North American Broadcast Teletext Specification, filed with the Federal Communications Commission, June 22, 1981.
6. Government of Canada, Department of Communications, Broadcast Specification. Television Broadcast Videotex, June 19, 1981.
7. The Bell System, Videotex Standard. Presentation Level Protocol, May 1981.
8. United Kingdom Teletext Industry Group, Petition for Rulemaking, filed with the Federal Communications Commission, March 26, 1981.
9. See, for example, A. B. Harris, Guide to Videotex Presentation Level Standards, British Telecom, July 1981.
10. CEPT Sub-Working Group CD/SE, Recommendation No. T/CD 6–1, European Interactive Videotex Service, Display Aspects and Transmission Coding; see also A. B. Harris, Guide to Videotex Presentation Level Standards, British Telecom, July 1981.
11. CBS, Inc., Reply by CBS, Inc., to Statements Concerning Petition for Rulemaking of the United Kingdom Teletext Industry Group, filed with the Federal Communications Commission, July 21, 1981, p. 8.
12. Lopinto, John, The Application of DRCS Within the North American Broadcast Teletext Specification, Time, Inc., no date.
13. Newman, William F., and Sproul, Robert F., *Principles of Interactive Computer Graphics,* McGraw-Hill, New York, 1973, p. 367.
14. Bown, H. G., O'Brien, C. D., Sawchuk, W., and Storey, J. R., A Canadian Proposal for Videotex Systems—General Description, Department of Communications (Canada), November 1978.
15. See Alternate Media Center, New York University, Access and Reception Quality in the Field Trial in Washington, D.C., Research on Broadcast Teletext, Working Paper Number One, September 1981; also Early Use of Graphics in the Field Trial in Washington, D.C., Research on Broadcast Teletext, Working Paper Number Three, December 1981.
16. Champness, Brian G., and Alberdi, Marco de, Measuring Subjective Reactions to Teletext Page Design, Alternate Media Center, New York University, September 1981.
17. Harris, A. B., Guide to Videotex Presentation Level Standards, British Telecom, July 1981.

18. United Kingdom Teletext Industry Group, Petition for Rulemaking, filed with the Federal Communications Commission, March 26, 1981.

19. Ibid., pp. 8ff.

20. U.K. Teletext System, 625 Line 50 Field System Applications, no date.

21. See, for example, W. S. Ciciora, Virtext and Virdata: Adventures in Vertical Interval Signalling, *Cable '81 Technical Papers,* National Cable Television Association, Washington, D.C., 1981, p. 102.

22. Telecaption—For the Deaf and Hard of Hearing. Training Manual, Sears Roebuck and Company, Chicago, 1980. See also John Lentz et al., Television Captioning for the Deaf, Signal and Display Specifications, Report no. E 7709–C, Engineering and Technical Operations Department, Public Broadcasting Service, May 1980 (revised).

23. NHK Research Laboratories, Japanese Teletext, March 1981, *BTS/Teletext Subcommittee, Interim Report,* vol. II, Electronic Industries Association, Washington, D.C., 1981, pp. 246–253.

24. *BTS/Teletext Subcommittee, Interim Report,* vol. I, Electronic Industries Association, Washington, D.C., 1981, p. 26.

25. Separate Statement of Commissioner Mimi Weyforth Dawson re Authorization of Transmission of Teletext by TV Stations, Federal Communications Commission, October 23, 1981.

26. Alternate Media Center, New York University, Access Time and Reception Quality in the Field Trial in Washington, D.C., Research on Broadcast Teletext, Working Paper Number One, September 1981, p. 28.

27. This observation was suggested by Professor Thomas Martin, School of Information Studies, Syracuse University, 1981.

28. Task Force D—Time Domain Equalizer, Interim Report, *BTS/Teletext Subcommittee, Interim Report,* vol. I, Electronic Industries Association, Washington, D.C., 1981, pp. 120–121.

29. Bernard Lechner, Efficient Utilization of the Broadcast Videotex (Teletext) Channel and Trade-offs with Other Issues Facing the Electronic Industries Association's Teletext Subcommittee's Standards Making Efforts, speech at Viewtext 82, New York, April 13–15, 1982.

6

Videotex, IR, and Other Things

As the foreword briefly described, the subject of teletext is often associated with videotex and is even considered to be a subset of the broad definition of videotex. (In the latter usage, "videotex" without a final "t" is synonymous with "videotext.") But what is the meaning of "videotex" and what does the term imply? The answer is that the meaning is still a little vague and it is best to recognize that so-called videotex systems are themselves a subset of two related larger activities known as "information retrieval," or IR, and electronic transaction or message systems. This chapter relates teletext to videotex, and videotex to computerized information retrieval and transaction systems in order to highlight the rather singular position of teletext within the whole evolving environment of computer/telecommunications.

That the words videotex and teletext are often associated with each other more or less mandates at least a summary of the videotex side of the matter. This is especially necessary when "videotex" is used, not as a broader category to which teletext belongs as a subset, but to distinguish two-way telephone-based systems from the essentially one-way broadcast systems. In practice, it has been far easier to define videotex by pointing to examples than by attempting to specify the distinguishing elements that identify a system as videotex rather than something else, because the generally stated distinguishing elements are almost always properties of many computerized information retrieval systems that are not called videotex.

Therefore, if we accept that the definition of videotex is still evolving, it seems reasonable to start with a historic approach to describing videotex. Subsequently, the collection of current videotex systems can be located within the world of on-line computer systems. Teletext, then, as described in this book, can thus be seen as a subject in its own right—a relatively new style of information/entertainment delivery that is indeed unlike previous computerized systems.

Viewdata to Videotex

In the early part of the 1970s, two separate activities occurring in England have since acquired at least one—and sometimes more than one—of the appellations "teletext," "viewdata," "videotex," and "videotext." It is unclear when the name "teletext" was first applied, but at least by 1972, when the BBC announced their Ceefax service, it was accepted that teletext was a generic term describing the technique of inserting digital pulses (representing "pages" or screens of text and graphics) into a normal broadcast television signal. The slightly later development by the British telephone company of an interactive computerized information retrieval system, using the telephone network to connect home television sets to the computer, was dubbed "Viewdata." In fact, the term Viewdata was intended to be the actual name of the service (i.e., a registered trade name).

There was then a somewhat simple distinction between teletext, a generic name for which the Ceefax service was the primary example, and Viewdata, the specific title of a proposed telephone company service. On the other hand, there did not seem to be a generic term for the British telephone company's Viewdata system, and it appears that the term teletext was sometimes employed to cover that system as well. A 1976 article in the United States, for example, used the word teletext to refer not only to Ceefax and Oracle but to the Viewdata service as well [1].

Unfortunately for the future of Viewdata as the title of one organization's proprietary system and service, the need for a generic name for telephone-based retrieval systems that looked like the British telephone company's own system caused the word "viewdata" to be applied in a more general sense. Ultimately, in 1978, the British Post Office (at that time the telephone side of the organization, British Telecom, had not yet separated from the Post Office) chose a new name for its proprietary system because the attempt to register the name Viewdata failed in England on the grounds that the term was really composed of two words in common use. The new name was Prestel, and it was once again a fairly simple matter to distinguish between teletext (exemplified by Ceefax and Oracle) and viewdata (exemplified by Prestel).

But the problem with terminology did not stop there. As countries other than England began developing their own versions of Prestel, there was a reluctance to accept viewdata as the generic term, and the word "videotex" began to be used as a substitute. In some people's minds, the word viewdata implied that the speaker was talking about a system very similar to the British system, whereas the systems developed in some other countries were enough unlike Prestel to warrant a different generic term. It may be instructive to note that the proceedings of the "first world conference" on viewdata, videotex, and teletext in 1980 used the word "viewdata" for the section on systems in Great Britain and the word "videotex" for the sections on similar systems in Canada, France, West Germany, the United States, and Japan [2].

During the late 1970s and early 1980s, the term videotex became more and more popular, until it is now used to describe not only Prestel-like systems but also the distinctly different teletext systems and a whole range of hybrid systems. Pushing it to the limit, some writers and speakers use "videotex" to encompass any and all of the computerized systems lumped under the similarly vague phrases "office of the future," "automated office," "electronic home," "information or communications revolution," and the like.

Stepping back from that nebulous state, we can begin a summary of videotex with a review of the "traditional" videotex systems in Western Europe and North America.

European Videotex

In 1981, when it became possible to speak of a European videotex system, it was an indication that one major stage of videotex development had ended and another had begun. A unified picture of European videotex began to emerge, replacing the previous image of antagonistic national videotex systems at odds with each other. The turning point came in mid-1981 when the 26 member nations of CEPT, the European Conference for Posts and Telecommunications Administrations, agreed to a standard for a basic level of videotex that incorporated rival British and French systems, as well as a number of new features. By mid-1982, at least a dozen countries in Western Europe claimed to have, or to be in the process of having, a public videotex service that could be accessed by any CEPT-standard terminal. As Table 6.1 indicates, however, only Great Britain, West Germany, France, and the Netherlands had any sizable number of registered videotex users in 1982, and in each case virtually none of the subscribers was using a CEPT-standard videotex terminal.

Therefore even though the trend in Europe is toward a unified public videotex system using a common standard or protocol, the situation in the early 1980s was still one of small individual videotex services. Each of the systems in the three major countries involved—Great Britain, West Germany, and France—are distinctive, giving at least that many different dimensions to the overall meaning of the term videotex.

As has been mentioned, Great Britain's Prestel more or less began the viewdata/videotex phenomenon. Several prototype Viewdata systems were developed during the period 1972–1975 by members of the British Post Office's telephone research laboratories. A number of influences were behind the development including research on

Table 6.1. Public Videotex Systems in Western Europe

Country	Service name	Subscribers
Austria	Bildschirmtext	Not yet available
Denmark	Teledata	Not yet available
Finland[a]	Telset	400
France[b]	Teletel, etc.	5,000
Italy	Videotel	Not yet available
Netherlands	Viditel	4,700
Norway	Teledata	100
Spain	Videotex	400
Sweden	DataVision	100
Switzerland	Videotex	Not yet available
United Kingdom[c]	Prestel	17,200
West Germany	Bildschirmtext	7,765

[a]Several names are used for different public services.
[b]There are two primary videotex trials and nearly 20 small pilot projects.
[c]Private systems are also in use, the most notable example being the London Stock Exchange's own videotex system called Topic, which has several thousand users.

bringing the "computer utility" to homes, research on the British equivalent of Picturephone as it was then, and a desire to encourage increased use of the telephone network.

Following a small two-year trial of the Viewdata system during 1976–1978, the British Post Office decided to move ahead with a more formal Test Service in 1978. (It was also in 1978 that Viewdata became Prestel.) However, basic decisions about how the Prestel service would be structured had already been made by the Post Office as early as 1975 [3]. For example, having concluded that the Prestel terminal should be a modified television set, the Post Office decided not to manufacture videotex terminals but to rely on the same tactic that the teletext broadcasters in England did, namely, encouraging television set manufacturers to produce the necessary receiver hardware. In the view of the Post Office's role as a public utility, the Post Office decided neither to provide nor to control the information to be delivered via the Prestel service. Therefore information providers or suppliers, who would create and store their own "pages" on Prestel's minicomputers, had to be cultivated as well.

Books, reports, articles, and papers have been written about the Prestel service and what it was supposed to be, what it should have been, and what it might be. In the United States, it has often been said that Prestel has failed, yet since its inception as a public service, Prestel has continued to grow (see Figure 6.1). Part of the reason for a misconception about Prestel's supposed failure has been due to some exaggerated claims made early in the game about the total number of subscribers that Prestel would have in a few short years. While these claims were apparently taken more seriously than they should have been, a brief discussion of what Prestel started out to be, and how it has developed, can aid in understanding what videotex is all about.

A former British Post Office official said in 1979 that the Post Office originally set out to create a universal computerized information retrieval system to be used by business professionals and consumers in the home all over the country [4]. That was what videotex was to be, and it is probably true to say that that elusive goal is still being pursued in country after country by numerous organizations. In reality, the Prestel system that emerged gave videotex a somewhat different image.

For one thing, the Prestel software (i.e., the set of computer programs that control the system) was written to maximize mass usage with fast response time at the expense of real processing power available to the user. So while the Prestel system could provide

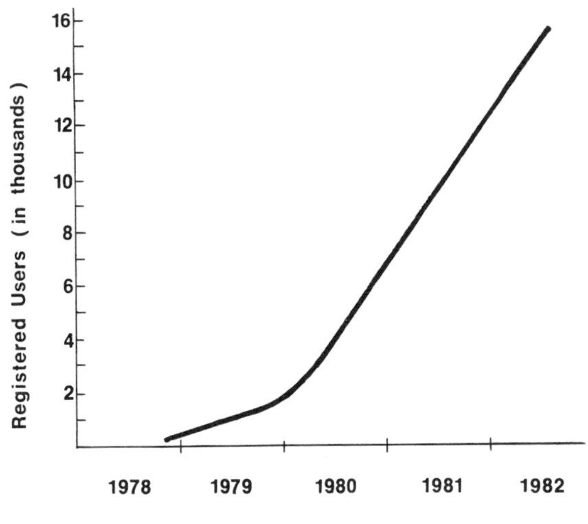

Figure 6.1 Growth in number of registered Prestel users.

a user with a requested page with no perceptible delay, that was in fact about all it would do. Subscribers could not use Prestel to calculate, write their own programs, play games that would involve anything other than page selection, or even send a message to another subscriber. (Messages to information providers were possible though.) And when simply seeking information in the Prestel data base, users could not type in a key word or subject term; information could be found only by making a series of choices or menu selections, or by going directly to any known page number. On the one hand, this provided a very simple and easy to grasp procedure for new users to learn, but alternatively, information seeking or transaction processing was considerably curtailed.

Another videotex image that developed directly from Prestel was that of ''information providers.'' The British Post Office's common carrier attitude, and a belief that the more people providing information, the wider the potential custom base, led to the solicitation of information providers. Because of pricing decisions made by the post office, these information providers had to pay an annual fee of about $5,000 to $8,000, an annual frame or page rental fee of $5 to $8 per page, and their own labor and equipment costs. Each information provider would have to access the Prestel minicomputers in order to create each page, with the Prestel software controlling the page creation process and the minicomputer's disks containing the stored pages. In something of a mixed blessing, Prestel did attract a large number of organizations that wanted to be information providers—hundreds initially, and nearly a thousand in 1982. (In Prestel terms, a true information provider is one who rents pages from British Telecom; a ''subinformation provider'' is one who subleases pages from an information provider. The total group of information suppliers in Great Britain is made up of about one-fifth information providers and four-fifths subinformation providers.) The problem was that the Prestel data base as a whole quickly began to appear as not very cohesive, terribly lacking in logical structure, and full of pages of questionable interest to anyone, created and then abandoned because the information provider did not want to incur the added expense of continually working on the pages.

Even before the public Prestel service began, it was apparent to some of the trial service participants that for economic, technical, and organizational reasons, the service was moving toward a business market rather than the consumer mass market [5]. Conceptually, this would present a division between what Prestel was designed for and what it was becoming: it looked like a mass market item, while it lacked the sophisticated processing capability of a business information service. The Prestel service would have lacked direction as well in the business market, if it had not been for the travel industry, which saw Prestel as a way to introduce at least some measure of computerization into the large number of small travel agencies that could not afford access to more powerful computerized reservation and information systems.

In the three years since the Prestel service began commercial operation, British Telecom has continued to adjust both technical and organizational aspects. Some of the technical additions include the ''mailbox'' facility so that customers can send electronic messages to other customers, and the ''gateway'' facility so that the Prestel system becomes essentially a conduit between a Prestel terminal and another non-Prestel host computer, which may in fact contain the sophisticated processing power suitable for business applications.

Perhaps more important for the Prestel idea of a universal information service, there has been a steady rise in the number of private Prestel-compatible systems or services in England. These systems are produced and sold by a variety of companies, including the computer giant IBM. The companies or organizations that have installed Prestel-compatible systems include such prestigious groups as the London Stock Exchange. These private systems, while technically compatible with Prestel transmission and display standards, often contain enhancements that are not part of British Telecom's Prestel. Over the next ten years or so, if current trends continue, it seems possible that private Prestel-compatible systems will substantially outdistance Prestel itself in both size and capability and draw data bases away from the telephone company's computers. Ultimately, British Telecom may find that the Prestel idea is indeed popular, but that its own Prestel service is used only for one or two specialized applications, such as directories and yellow pages. In other words, the telephone company's own Prestel computers will store primarily the traditional telephone company information products.

Not surprisingly, France, another country that has helped to shape the meaning of "videotex," actually began with the idea that the telephone company's videotex service should be essentially a conduit for information stored on private computers, and that the telephone company's own data bases would be limited to directories [6]. French videotex planners in the late 1970s had concluded then that the videotex system would be a communications link to be exploited by various organizations, with their own particular application being the telephone directory. Also not surprisingly, French officials, like their British counterparts, found it easy to succumb to making grand projections about the success of the videotex service. Early forecasts were that 30 million videotex terminals (i.e., small telephone units with a built-in black and white screen, as in Figure 6.2) would be distributed free of charge by the telephone authorities between 1983 and 1992, to replace completely the paper telephone directories [7]. However, the early years of the electronic directory project do not suggest that such numbers will be reached. Original plans called for the electronic directory service to be given a full-scale test in 1982 with 270,000 terminals in the hands of telephone subscribers in the Ille-et-Vilaine region. Yet by the latter half of 1982, fewer than 2,000 terminals were in place in Ille-et-Vilaine.

Figure 6.2. Teletel terminal. *(Courtesy of Intelmatique.)*

One of the aspects of the French videotex experience that is pertinent to understanding what videotex is about is the French acceptance of the videotex idea as part of a broader concept called "telematique" (or "telematics"), which is another name for the merging of computers and telecommunications. The government-sponsored Telematique Program dates back to a 1975 decision to upgrade radically telephone service in France. As the program developed it has come to include not only the electronic directory project and the projected national videotex network called "Teletel," but also facsimile equipment, electronic blackboards, audio teleconferencing, and the "smart card," or microcircuit card—a piece of plastic that looks like an ordinary credit card but contains a build-in microprocessor and a certain amount of memory.

In essence, in France, videotex is only one element of the telematique realm, but a confusion still exists in the United States over which of several names to apply to specific French systems or services. For example, "Antiope" is one of the best known names (in the United States) of the Telematique family because it is used for both the initial French teletext service and for the common display standard adopted for French videotex and teletext. Unfortunately, outside of France, the term Antiope is often used to refer to all of the French videotex and teletext efforts and even to the Telematique Program as a whole. Following from that, an equally broad definition is given to the word "videotex" when applied to the French experience because "Antiope" is considered (incorrectly, of course) to be the term for the whole of the French version of videotex.

Even in cases where the terms Telematique (the overall program), Antiope (the display standard and the teletext service), Didon (the transmission protocol), and Teletel (the national videotex system) are used correctly, there is a definite difference between the original British idea of videotex as incorporating a color television set and the French idea of videotex, where the largest application would involve a terminal that is a telephone-video screen combination, lacking color and the ability to receive television.

Although the development of videotex (Teletel) in France remains behind that of Prestel in Great Britain, a similar trend toward special purpose systems is apparent, spurred in part by the initial decisions to create Teletel as primarily a communications link between customers and private or third party data bases. A mid-1982 compilation of videotex in France lists the electronic directory trial, a separate Teletel trial, and some 19 specialized trial services (see Table 6.2). Almost all the special trial services have fewer than 100 participants. Some of the experimental services are actually parts of a larger effort (e.g., Telagri, Agrinfo, and Telecoop are part of an effort to use videotex for agribusiness purposes). Telagri will provide agricultural management and technical information, Agrinfo will provide weather forecasts and specific advice for farmers, and Telecoop will be designed for members of farm cooperatives. Other experimental services are of the stand alone variety, such as Cititel, a travelers' or tourists' information service available at Teletel terminals in public locations such as hotels and street corner kiosks.

Another Western European country that has had a major role in creating the videotex mystique is West Germany. Although West Germany did not spawn its own standard for the transmission and display of videotex codes, as Great Britain and France did, West Germany did influence the development of the pan-European CEPT agreement on videotex standards and is in the process of establishing a complex national videotex network/service called Bildschirmtext.

West Germany did not begin a formal videotex trial until 1979–1980, but by that time a consensus had already developed regarding the meaning of videotex as a tech-

Table 6.2. Videotex Pilot Projects in France

Project name	Number of users
ADM (medical applications)	200
Agrinfo	100
Ares	Under 50
Cititel	100
Claire	Under 50
Credit Agricole	Under 50
Electronic Directory	1,500
Family Allowance Regional Service	Under 50
Metronic	Under 50
Ministry of Education	Under 50
Nantes CRCI	Under 50
Nouvelles Frontières	Under 50
Social Security Regional Service	Under 50
Telagri	Under 50
Telecoop	Under 50
Telem	Under 50
Teletel (3V)	2,500
Telinformation	Under 50
Todel	Under 50
Videobanque	335
Viniprix	Under 50

Source: Based on Gary H. Arlen, *International Videotex Teletext News*, June 28, 1982, p. 9.

nology for all forms of data communication and the subsequent need for enhancements beyond the videotex capabilities as defined by Prestel. As an example, the Germans felt that a useful videotex standard required a character set with more characters, such as characters with diacritics, than those available in the British set, and with more graphic "characters." This, and a decision to connect various attributes to a single character, led the Germans to accept the French suggestion that at least 16 bits be allotted per character—eight bits to define the character and the remaining eight bits to indicate attributes [8]. In 1981, the CEPT agreement, strongly supported by West Germany, brought together the British and the French approaches to relating attributes to characters by incorporating both approaches.

The West German videotex system is also often credited with the development of the videotex gateway, even though the French similarly saw the national network as essentially a gateway to private third party data bases. In West Germany, however, owing to a political and regulatory environment that is very careful to protect the rights of individuals with regard to information stored in distant data bases, there has been much more active development of the gateway concept, so that the Bildschirmtext central computers are actually intermediaries between customers and the computers of banks, retailers of various kinds, publishers of magazines and newspapers, and the like.

The first trials of the Bildschirmtext system (or BTX, as it is called) began in 1977 after the German telephone authority, the Bundespost, purchased a copy of the Prestel system from British Telecom in 1976. At about the time that the 1977 pilot trial began, the Bundespost also hired a British company, Systems Designers Ltd., to develop the software for interconnecting the videotex processors with external or private computers.

This software package, sometimes called the External Communications Network and at other times simply "the gateway software," is currently used by Bildschirmtext, and a copy was also sold by the Bundespost to British Telecom in 1980 so that Prestel could begin adding the gateway feature.

The early West German pilot trials have evolved into a larger, formal trial involving nearly 8,000 users who can dial to the Bildschirmtext central computers in Berlin and Düsseldorf. These central computers are in turn connected to about 20 external computers, with the data communications between the Bildschirmtext computers and the external computers traveling via the national packet-switched data network, Datex-P. During 1983, Bildschirmtext is scheduled to grow from a trial service to a public commercial service. The West Germans, too, have great expectations for their videotex system, anticipating a million users by 1986.

One of the more attention-getting aspects of the Bildschirmtext arrangement is the interconnection of home consumers with their banks via the videotex system. One West German bank, Verbraucherbank, claims to be particularly successful in "home banking" through the videotex system. However, Verbraucherbank had begun to computerize its operations and to install self-service banking machines in 1977, prior to using videotex, and much of the bank's success in going from 3,000 to 41,000 accounts over the past six years has been due to the attractively priced services made possible by generally automating as much as possible. The relatively few bank customers who use Bildschirmtext to reach the bank's computer benefit from trial-level prices for videotex equipment and service, and the generally favorable response by these customers may not necessarily continue when Bildschirmtext becomes a commercial service.

In general, Bildschirmtext brings to the overall definition of videotex the image of a national network linking customers to the whole gamut of external computers and services, including home banking.

The European experience, then, in shaping the meaning of videotex began with a government telephone company installing a computer system linked to modified television sets in order to provide pages of information. Videotex then evolved in a dozen or so European countries into an information system, a communications system, a display standard for generating characters and graphics on video screens, and a procedure for mediating between certain types of terminals and virtually any digitized service. And while it is easy to speak of the British concept, or the French contributions, or the West German plan, that is certainly an oversimplification; many individuals and organizations in a number of countries have contributed to the overall result by developing one feature or another. Yet in Europe, it is nonetheless relatively easy to point to an example of a videotex system—Prestel, Teletel, Bildschirmtext, and so on—and such systems are almost all operated by the national telecommunications authority as a new improvement to the national telephone or data communications network (i.e., as a way to provide access to computerized data bases and to computerized transaction services).

On-line Industry

In the United States, however, and to a lesser extent in Canada, it has not at all been easy to say with precision what videotex is, because computerized information services existed on a national scale before people began to talk about viewdata or videotex. Where is the dividing line between a computerized information system used at home with a microcomputer as the terminal and a television set as the display device, and a so-called videotex system? Table 6.3 lists some representative on-line information ser-

Table 6.3. National On-line Systems Based in the United States (a representative list)

Service name	Number of subscribers
Dow Jones News/Retrieval Service[a]	45,000
Compuserve Information Service[a]	26,800
The Source (Reader's Digest)[a]	20,000
Control Data Corporation's Cybernet	17,300
Lockheed's Dialog	15,000
CSC's Infonet	9,500
General Electric Information Services	6,300
System Development Corporation's Orbit	6,000
Reuters Monitor	4,500
Telerate	3,000
Mead Data Central's Lexis and Nexis	3,000
New York Times Information Bank	2,600
Medline	2,000

Source: Based on Online Database Growth Continues, *IDP Report* 2 (23): 4 (February 5, 1982); and other sources.

[a]Denotes services that have been called "videotex."

vices available nationally in the United States, and internationally, via packet-switched data networks. When videotex became a popular topic in the late 1970s, at least three of these services—the Dow Jones News/Retrieval Service, Compuserve, and The Source—were often granted the status of "videotex," presumably because these services were oriented toward the mass market, even though they did not utilize color and graphics and the average terminal was not a home television set.

In fact, prior to the popularization of the term videotex in the United States, the umbrella term for computerized information services was "the on-line industry" or "the data base industry." In the very early 1970s, before videotex really existed, there were an estimated 150 on-line systems in the United States and about 20 major information retrieval systems or services in use or on trial [9]. Some of these systems have now been around for a decade or so (e.g., Medline began in 1971 based on a 1970 predecessor, a Mead Data Central retrieval system was in use in 1969–1970, the New York Times Information Bank began as an in-house service in 1969–1970, and the original Dialog system was operational in 1966). It is worth noting that while European videotex systems are often stigmatized as "government funded," the on-line systems and services developed in the United States in the late 1960s and early 1970s were also heavily funded by the federal government in the form of R&D contracts.

One of the on-line services that has acquired the label of videotex even though it predates "normal" videotex and does not (yet) employ color and graphics is the Dow Jones News/Retrieval Service. Beginning with a joint venture with Bunker Ramo in 1972, which eventually became the Dow Jones News/Retrieval Service in 1977, Dow Jones has sought an ever wider audience for on-line information. Recognizing in the late 1970s that new subscribers might be people who already owned microcomputers, Dow Jones began actively to work with Apple Computer, Inc., to promote the use of Apple microcomputers as terminals for interacting with the on-line news and information service. Similarly, Dow Jones has been involved in several cable television adaptations of their information service (including the use of the teletext technique, as well as two-way digital techniques) and has created other adaptations for access via Great Britain's Prestel service and Knight-Ridder's Viewtron service (the latter is a videotex service on trial currently in Florida). While prior to 1980 the News/Retrieval Service carried only information generated by Dow Jones itself, the company has begun to add data bases from other information suppliers. The number of users of the service has risen from 5,000 at the end of 1978 to nearly 50,000 in the latter part of 1982, with a growth rate of about 2,000 additional subscribers per month (see Figure 6.3).

A sweeping view of the development of the on-line industry in the United States seems to show that the perceived need for easier access to information led to research in computer processing and software design, searching for faster and yet more complex ways to process requests, or interrogations, from a user. The procedures that the customers could employ (i.e., the computer commands or search statements that could be used to find items of information in a data base) became rather complex themselves and almost assuredly differed from system to system. Part of the videotex concept, then, as developed in Europe, was to answer this problem by replacing the complexity with very simple procedures—limiting the problems for users but also curbing the power of the systems.

When videotex arrived in the United States, the label was not only applied to a few of the easier to use on-line services but also to some new systems that really did look like the videotex systems in Europe (and in Canada and Japan). Some of these new systems used the software developed in Europe and Canada, and some used homegrown varieties. Radio Shack (a division of Tandy Corporation), for example, has its own videotex system that has been used by a number of organizations for relatively small applications, while AT&T has also developed a videotex system, presumably for very large-scale applications. Just keeping track of all the trials of videotex systems and services has spawned its own miniindustry of reports and newsletters, which list several dozen videotex trial services in the United States (see Table 6.4).

One of the factors affecting the entrance of videotex as such into the United States, mentioned in previous chapters and noted in Table 6.4, has been the cable television

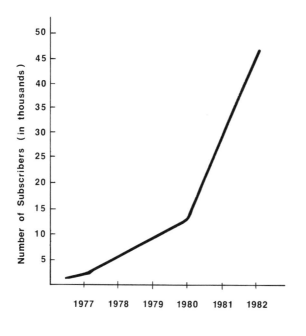

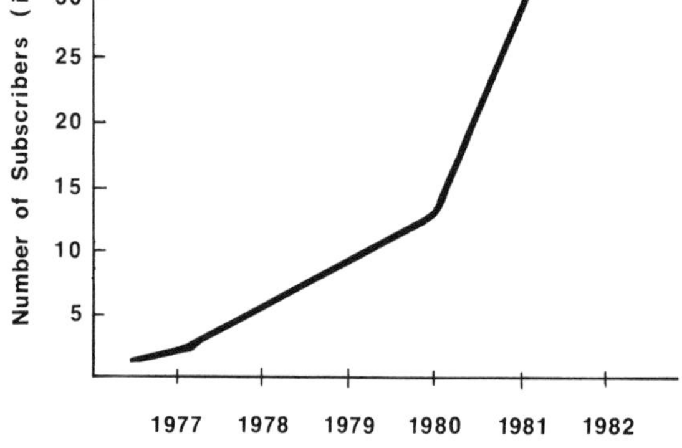

Figure 6.3. Growth in number of Dow Jones on-line users.

Table 6.4. Videotex Trials In the United States

Service name
Advertiser-Tribune Electronic Newspaper (Ohio)
Bank-at-Home (Tennesee)
Bison (Texas)[a]
ConTelVision (Virginia)
Day and Night Video Banking (California)
Electronic Editions (Washington)
Firsthand (Minnesota)
Greenthumb (Kentucky)[a]
Harris Electronic News (Kansas)
HVC Corporation (Texas)
Indax (California, Nebraska)[b]
Instant Update (Iowa)
Project Pronto (New York)
StarText (Texas)
Times Mirror Videotex (California)[c]
Venture One (New Jersey)
Viewcom/Grassroots (California)
ViewTimes (Connecticut)[b]
Viewtron (Florida)

[a]Currently suspended.
[b]Uses cable television systems.
[c]Uses both telephone lines and cable television system.

industry. Some of the cable's text services resemble videotex systems very closely because the pages of text are created using one of the videotex standards for creating text and graphics. And on the cable systems where true interaction can take place between a subscriber terminal and a distant computer, there may be really no difference between that service and a telephone-line-based videotex service. In fact, whether videotex will be more successful as a part of cable television or as part of the telephone network is a vigorously debated question. Critics of the telephone's capabilities claim that the telephone network has not been constructed to handle large numbers of lengthy, simultaneous calls to computers, and that the network will become overloaded to the breaking point. Critics of the cable's potential point out that most cable systems already built cannot handle two-way digital communication easily, and that cable television systems are also structurally unsuited for large-scale interactive digital communication. (Again, as mentioned in earlier chapters, cable systems can use the teletext technique to digitally provide pages of text and graphics that answer many information needs without the requirement for a digital channel from each subscriber back to the computer.)

ANSI Videotex

In the end, unless "videotex" assumes an all-encompassing connotation, the definition of videotex may depend upon the existence of an accepted standard for the display of videotex information. In the United States this is likely to be the standard (mentioned in previous chapters) published by the American National Standards Institute (ANSI), called the *Videotex/Teletext Presentation Level Protocol Syntax (North American PLPS)* [10]. A system that has the characteristics described in the North American PLPS, or similar characteristics, will be classified as videotex. Although it is beyond the scope of this chapter to describe the North American PLPS in detail, it is worth summarizing the features of the system as growing out of the Canadian Telidon system, the European systems (both Prestel and Antiope/Teletel), and the AT&T system.

The Canadian Telidon system dates back to work in the late 1960s at the Canadian Department of Communications' Research Centre on computer graphics techniques [11]. From 1969 to 1976, the Canadian researchers defined a set of graphic commands, or Picture Description Instructions, that could be transmitted to a terminal to cause a graphic design to be drawn. The use of graphic commands is a fairly common practice in the computer graphics world, and the Canadian PDIs were one solution to the definition and implementation of graphic commands. Partly in response to the mosaic graphics of Prestel and Antiope/Teletel, the Communications Research Centre announced its own videotex system, Telidon, in 1978, based on the exclusive use of PDIs to generate graphics.

The attractiveness of the Telidon graphics led to purchases of the Telidon system by several companies in the United States, including Time, Inc., and the Times-Mirror Company, and to a string of videotex trails in Canada (see Table 6.5).

The Canadian PDIs were also attractive to AT&T, when researchers at Bell Labs began working on a videotex standard for the Bell System. The AT&T standard, published in mid-1981 as a "Presentation Level Protocol," used the PDI structure to incorporate the mosaic graphics and other features of Prestel and Antiope, as well as the graphic commands or geometric graphics. For example, while the original Telidon specification speaks of eight PDIs with one of those being CONTROL, the AT&T definition listed the PDI set as containing two major subsets (six graphic commands and eight control codes), while apart from the PDI set are the TEXT sets that include the mosaic graphics. In the AT&T scheme, the control codes (which are a subset of the PDI set) are used to set the color and other attributes of the TEXT sets.

In early 1982, after considerable discussion of whose standard was best, the matter came under review by a subcommittee of the American National Standards Institute. The subcommittee took the AT&T–defined system as a starting point and worked

Table 6.5. Videotex Trials in Canada[a]

Service Name
BC Tel (British Columbia)
Cantel (national)
Edimedia (Quebec)
Elie (Manitoba)
Fishnet (New Brunswick)
Grassroots (Manitoba)
Ida (Manitoba)
Lepage Ltd. Real Estate Listings (Ontario)
Mercury (New Brunswick)
Newfoundland Telephone's Tourist Information
Novatex (international)
Saskatchewan Wheat Pool (Saskatchewan)
Teleguide (Quebec)
Videotron Intervision (Quebec)
Vidon (Alberta)
Vista (Ontario, Quebec)

[a]There are at least 50 trials of Telidon videotex in progress or about to begin in Canada. This is a representative list.

jointly with a similar Canadian standards committee to arrive at a videotex standard that would be acceptable in both the United States and Canada. This new document, the *Videotex/Teletext Presentation Level Protocol Syntax,* will go (during 1983) through the formal channels for adoption by the American National Standards Institute as a voluntary industry standard.

Thus equipment and systems that conform to the ANSI videotex standard could be properly called "videotex." But one thing that has become clear is that so-called videotex systems, using just about anybody's standard, are quite capable of being modified, enhanced, expanded, and so forth, as has happened to Prestel, Teletel, Telidon, and others. And videotex systems have changed functionally as well as technically, whether individually or as a whole. The trend seems to be that videotex systems are losing their simplicity and becoming more like the familiar on-line systems, while the on-line systems are adding color, graphics, and menu selection, and consequently looking more like videotex. If so, then there seems to be no relief in sight for resolving the fuzziness of the definition of videotex—it will still connote some broad category of computerized information/entertainment/transaction systems that may or may not be mass market-oriented and may or may not use color and graphics.

Videotex, in sum, is a part of the on-line industry, the development of computerized information/transaction systems, and can be seen as both a step backward (in information retrieval sophistication) and a step forward (in adaptation for wider use). Teletext, however, or more properly the use of the teletext technique, still stands out as not quite like anything previously available. We have had on-line retrieval systems for nearly three decades, but we have never really had a whole new digital channel hidden within each of our broadcast or cable television channels. Eventually, writers may decide that use of the teletext technique is a minor subset of videotex, but the fact remains that the teletext technique carries its own vast and particular potential for new information/entertainment services and transmission procedures.

References

1. Hill, I. William, Seek Editor Input on Future of Video Screen Newspapers, *Editor & Publisher,* September 4, 1976, p. 14.
2. *Viewdata 80, First World Conference on Viewdata, Videotex and Teletext,* Online Conferences, Ltd., Northwoods Hills, England, 1980, pp. v–vi.
3. Woolfe, Roger, *Videotex,* Heyden & Son, Ltd., London, 1980, pp. 74–76.
4. Bright, Roy D., Prestel, the World's First Public Viewdata Service, *IEEE Transactions on Consumer Electronics,* CE–25(3): 252 (July 1979).
5. See, for example, Woolfe, *Videotex,* p. 75.
6. Bright, Roy D., The Telematique Program in France, *Viewdata 80,* Online Conferences, Ltd., Northwood Hills, England, 1980, pp. 19–21.
7. The French Connection, *Viewdata/Videotex Report,* January 1980, pp. 28–29.
8. Zimmermann, Rolf, Future Utilization of Interactive and Broadcast Videotex in Germany and Its Effects on Standardization, *Viewdata 80,* Online Conferences, Ltd., Northwood Hills, England 1980, pp. 264–266.
9. Lancaster, F. W., and Fayen, E. G., *Information Retrieval Online,* Melville Publishing Company, Los Angeles, 1973, pp. 63–64.
10. American National Standards Institute, Committee X3L2.1, *Videotex/Teletext Presentation Level Protocol Syntax (North American PLPS),* Draft Standard, June 18, 1982.
11. See, for example, Herbert G. Bown and William Sawchuk, Telidon—A Review, *IEEE Communications Magazine,* January 1981, pp. 22–28.

7

Futures for Teletext

The previous chapters have dwelt on details and descriptions of the teletext technique and services using the teletext technique. Now it is time to place the subject in perspective to see how teletext stands in relation to other means of disseminating information or entertainment electronically. Certainly, the technology is less important than the services for which it is employed—molded by the viewpoints of the service designers, the economic environment, politics, law, and the like. Moreover, it is almost axiomatic that a new technology will eventually assume a form and function not quite like what the original designers intended. The telephone, for example, was originally considered to be a device for delivering audio signals to group audiences in a one-way direction, like a wired radio service [1]. Of course, it did not quite turn out that way although the telephone network *is* used to carry radio and television signals from production centers to broadcasting stations, and also in applications such as tele-lecturing where the audio channel can indeed be one-way from a lecturer to a remote audience.

The purpose of this chapter is to add some context to the teletext discussion by examining relations with the past and the present, and projecting a future. Actually, the word "future" should be in the plural, because there will no doubt be more than one path followed by teletext developers. Beyond that, some commentators have ventured to discuss teletext as part of the entire scale of social history and human communication. It is interesting to speculate on the centuries-long effects of the use of teletextlike systems as a force akin to television itself but with different parameters and different results. But first, a look backward.

Paralleling the Past

The first step in understanding the potential of a new technology is to recognize that it is probably not all that new, at least in function, and that parallels do exist in the past. The second step, however, is to realize that the parallels are rarely if ever exact. Some parts of the past must be ignored as irrelevant, and some parts must be accepted as good indications of the future when modified by an assessment of the current environment.

To understand how teletext might develop, we could look back to the early days of similar systems: radio, television, color television, cable television, and computerized information systems. Each of these is similar in some respects and different in others with regard to the current and near-future state of teletext.

Radio broadcasting (i.e., not point-to-point communications) started as such around 1910, and by 1916 at least one person, David Sarnoff, was proposing that regular programs be broadcast and that consequently hundreds of thousands, even millions, of radio receivers would soon be sold [2]. The development of the radio industry was soon slowed by World War I, but at the same time the technology was enhanced by research and development that took place as part of the war effort. In 1920, less than a handful of radio stations were broadcasting programs, but four years later over 500 stations were on the air and sales of radio receivers were bringing in roughly a quarter million dollars per day. As the 1930s began, the number of radio receivers in use exceeded 17 million. In short, as soon as the war ended, the growth of radio was dramatic.

Are there similarities with teletext, if the war years can be ignored? Both broadcast teletext and broadcast radio require that the listener/viewer purchase a piece of equipment to receive the signal. In 1920, the cost of a radio receiver, about $75, was not a negligible sum. Yet David Sarnoff estimated (after the war) that in one year 100,000 radios would sell at that price; in the second year, 300,000 radios would sell, and in the third year, 600,000 [3]. As it turned out, this time Sarnoff was right.

Both broadcast radio and broadcast teletext also face(d) a ''chicken and egg'' problem in stimulating demand. Presumably, services or programs do not develop unless there is an audience, but the audience is not likely to purchase receivers unless there is a reason (an available service) to do so.

Several other aspects of the development of radio broadcasting may also be useful to consider as indications of a similar growth (or lack thereof) for teletext. First of all, radio was dominated in the early days by an array of amateurs. It was not difficult to acquire the necessary pieces of hardware to build radio receivers and radio transmitters. In 1922, when there were over 500 broadcast stations, there were nearly 17,000 amateur radio stations. In fact, the efforts of amateurs to broadcast material for ''general listening'' was one of the major forces that led to the beginning of the radio industry as such.

A second aspect is the power and influence of the companies that had the most to gain from the success of radio broadcasting—the manufacturers of equipment, and especially the manufacturers of radio sets. During World War I the equipment manufacturers were able to consolidate their positions by buying or exchanging patents and patent rights. Soon thereafter these same manufacturers were the ones who dominated the airwaves [4].

As radio grew, what was it used for? Primarily, it seems, for entertainment, along with some news and sports features and to a lesser degree education, religious programs, and the electronic version of snake oil salesmen.

Each of these aspects has some implication for teletext and especially for estimating whether or not broadcast teletext might have a similar growth history. Regarding the influence of the amateur radio operators, some suggest that the same will be true for teletext (i.e., in the early years amateur builders of teletext receivers will provide the spark for growth) [5]. However, there are differences between the environment for radio amateurs in the 1920s and the environment for teletext amateurs today. First, the necessary electronic components are not readily available and inexpensive, and the receiver is not simple enough to be built by people with virtually no related knowledge. That is not to say that amateur teletext receivers cannot be built, but rather that there is a considerable difference in the scope of the undertaking. Second, the early radio amateurs could transmit almost as easily as they could receive, while teletext amateurs today have no access to the vertical blanking interval of a broadcast television signal except perhaps in the case of a few small educational stations. Thus amateurs are not likely to provide the incipient services. Consequently, to the extent that the existence of the amateur radio influence was critical to the early success of radio, we can suggest no such rapid success for teletext (ignoring for the moment other factors).

Another implication, arising from the influence of the manufacturing companies over the production of radio programming, is that teletext may yet have a chance for rapid success if the manufacturers of teletext–television sets become the sponsors and producers of the teletext services. These companies would have a strong financial motive for creating successful teletext services in any way that they could. But this is not the case, at least not at present, with teletext. Teletext has been born into an existing array of broadcasting companies, semiconductor manufacturers, computer companies, and electronic equipment manufacturers. The television stations that are broadcasting teletext are neither the builders of the teletext origination systems nor the manufacturers of teletext receivers. There is one, still small, exception though. The NBC television network intends to begin a national teletext service, and NBC's parent company, RCA, manufactures television sets and has been deeply involved in experiments with teletext receivers. On this matter, there may be some reason to believe that teletext could indeed parallel radio.

Finally, there is the recognition that early radio was primarily entertainment, and a considerable novelty. Teletext shows only a few signs of being an entertainment medium—broadcast teletext is often portrayed as a news and information service because of the way it was developed by the BBC in England—and there is certainly little new about seeing text and graphics on television screens.

Although this treatment has not covered all the possible similarities between early radio and early teletext (it would probably not be useful to push the comparison too far), there is reason to say that teletext will not be as successful as early radio if the elements for success are similar, because teletext fails to show several of the key elements found in radio. However, there is the exception of the relationship between equipment manufacturing and service production; there could easily be a similar influence on the growth of teletext. On the matter of early radio being primarily entertainment, teletext will probably be more successful (i.e., more widespread) if it is developed as an entertainment medium than as an information service [6].

Another technology we might match teletext against is television itself. In fact, teletext's relation to television could be very similar to television's relation to radio. In both cases, an existing broadcast service was well known, and the new broadcast service did not, in some people's minds, add much to the existing medium. And in both cases the receiver was not inexpensive and was not easily built at home.

The development of television broadcasting was unlike the development of radio. Television underwent a much longer gestation period that was slowed by technical disagreements and standards arguments before dramatic growth eventually occurred within a few years of the industry's actually getting started. Experimental television existed almost as early as experimental radio, and as early as 1915 Guglielmo Marconi was talking about the equivalent of Picturephone, a visible telephone [7]. Experimental television was broadcast over the air and transmitted by telephone lines and coaxial cables during the 1920s and 1930s. The first application for a commercial television license was not filed with the FCC though until 1939. For the next two years, engineering standards were discussed by the National Television System Committee (NTSC), leading to substantial industry consensus and subsequent authorization by the FCC in 1941. But the Second World War intervened and at least four of the ten existing television stations went off the air. When the war ended and television began again, previously unsuspected interference problems became apparent and the FCC ceased granting new television licenses in 1948 (between 1945 and 1948 the number of television stations had risen to just under 100). For four years, the FCC studied a range of television problems, including new engineering standards, technical systems for color television, and the allocation of the VHF and UHF channels under a new national plan for assignments. The "freeze" on television station licenses was lifted in 1952, and within five years over 500 stations were on the air. The number of television sets manufactured per year rose from 179,000 in 1947 to over 7 million in 1950, and now is about 17 million per year, including imports.

Obviously television had immediate consumer appeal, because it offered something new even if the general use of television paralleled the general use of radio (i.e., to broadcast entertainment, news, sports, weather, and all the other things we are familiar with). But several aspects of television's growth could presage a pattern to be followed by teletext. First, television did not really leave the experimental stage until after a two-year "standards debate" culminating in industrywide agreement on how to begin and in FCC permission to do so. There is a strong suspicion that, just as in the case of television, teletext will not develop widely without the foundation of industry consensus and authorized standards for technical implementation. Second, one of the forces behind television growth, after it did get going, was the existence of radio broadcasting organizations that developed television networks to match their radio networks. The existing networks may well be the same companies to provide network teletext. A third aspect of the television/teletext comparison is that delays caused by engineering difficulties, the war, and standards discussions apparently did not cause any serious damage to either the health of the industry or the social well-being of the populace. In fact, television critics may argue that the delays served to save the public, at least for a few years, from the disadvantages of television.

Again, though, one aspect of both radio and television is not true for teletext yet, and that is the obvious entertainment value.

Aside from the comparisons with radio and television as such, comparisons can also be made with certain additions to the television environment (which is what teletext is) such as color television and cable television.

In the case of color television, there appears to be a direct reflection of teletext in that both only add a new feature to television reception, both realistically require the user to purchase a new television set (although teletext can be received with a set-top adapter, reception quality will be inferior), and both did not have obvious benefits for consumers in the early days of the technology.

Color television was the subject of experimentation as early as the 1920s and was considered by the FCC in 1941, but it was not until after World War II, in 1946, that a company (CBS) petitioned the FCC to adopt a color standard. The FCC denied the petition on the grounds that more experimentation was needed, especially to create a color system that would use a bandwidth identical to the bandwidth of the black and white television signal. In 1949, three different techniques were proposed to the FCC by three different companies—CBS, RCA (parent of NBC), and Color Television, Inc. Although none of the systems met all the objectives that the FCC had in mind, the CBS system came closest and, after some initial hesitation, was declared the standard in 1950. But owing to a shortage of materials needed to produce color sets, the National Production Authority forbade the manufacturing of color televisions. Consequently, the standard did not really get established and by 1953 new technical developments permitted a new color system to be feasible. This new system was supported by the National Television System Committee and subsequently authorized by the FCC. The new standard solved one of the major problems of the CBS system and other earlier systems, namely, the reception of a color signal by a black and white set for display as a black and white picture.

Once color television broadcasting began about 1956, it took approximately ten years before evening network television programs were almost all in color. At about the same time, 1966–1967, the annual production of color television sets began to equal the annual production of black and white sets. In recent years the production of color sets has exceeded that of black and white sets, but not overwhelmingly so. Millions of black and white television sets are still manufactured every year.

The question of how color television got started during that decade from 1956 to 1966 can be answered by pointing to several influences. One certain influence was the self-interest of the television manufacturers (and the one network with a manufacturing relationship—NBC). Television set distributors and retailers, often in cooperation with a local station, were encouraged actively to promote color television in order to stimulate the purchase of color receivers. But another strong influence was a demonstrated tie between color and advertising, on two levels. For one thing, advertisers who sponsored programs in color knew that the viewers with color sets were a special audience segment—affluent, status conscious, apt to buy convenience products, and looking for new things [8]. In addition, audience research began to show that color commercials were twice as effective as black and white commercials. Advertisers, who knew that their production costs might rise by as much as half to add color, were nonetheless willing to spend the money to see the effectiveness of the advertising more than double. This relationship of color to advertising also gave an edge to stations broadcasting in color in areas where other stations stuck to black and white. The color station could dominate the market by attracting more advertising revenue.

Therefore, there are lessons to be learned from the example of color television. First, the push to introduce the technology is likely to come from companies that will benefit from set sales. Second, advertisers will know that, with the new technology, they will reach a more affluent audience with discoverable characteristics that make it easier to match products and services with potential consumers. But the third lesson is that without a more substantial link to advertising effectiveness, a major force behind market penetration is lacking. On the other hand, teletext can chart its path to success if that advertising potential can be exploited. Advertisers will turn to teletext, and broadcast stations will promote teletext in a two-pronged effort in which each feeds the other. Finally, even if the new technology is successful, it will probably take several decades to replace the existing universe of sets in use. Even though the great majority of all television broadcasting is in color, monochrome receivers are still in use and are still being manufactured and sold. It seems equally likely that, even after teletext begins in earnest, it will be a long time before *all* new television sets are sold with built-in teletext capability.

A second addition to the television environment to consider as a possible parallel to teletext is the introduction of cable television. The cable television industry dates back at least to 1948 when a television set retailer in Pennsylvania decided to erect his own high antenna to receive television signals in an otherwise nonreception area and to wire his own and others' sets to the antenna. For 20 years, cable television was just that—a retransmission facility to allow viewers to see channels that they would otherwise not see unless they erected their own antennas on a mountaintop or similar lofty location. As the 1970s began, cable systems were starting to be discussed as more than retransmission systems, and public attention was drawn to the interactive possibilities for broadband cable networks, ranging from simple meter-monitoring by a headend minicomputer to full two-way video communication. But the real change in cable television came in the middle and late 1970s when it was discovered that cable customers would pay additional fees for "premium programming," uncut movies, and special events, and that the premium programming could be delivered to a national audience by using communications satellites to reach the thousands of individual cable systems. More recently, cable is again being promoted as a vehicle for new services and new features, but the profit is still in basic retransmission and in premium programming or pay TV.

Generally, then, the first 20 years of cable were very utilitarian. Cable systems developed where they served a real need. In a way, this is already the case in the first few years of teletext. The teletext technique is being used commercially to multiplex digital information onto the vertical blanking intervals of superstations. This is less expensive than transmitting the same data by other means such as leased telephone lines. And because the data being transmitted are intended for the same receiver as for the television signal—the cable headend—it makes sense to put the two together. Another utilitarian aspect of teletext is in situations where there is a logical link between the video program and the digital data in the vertical blanking interval. The most obvious example is closed captioning. But also in new services such as the national weather channel for cable television, the digital stream is used to provide program-related information (e.g., additional weather information sorted and tagged for individual geographic areas) and to continue the weather information when the video program switches each evening to pay TV for a few hours. Thus we might see a period of teletext growth where the utilitarian nature of teletext as a relatively inexpensive transmission facility is the basis for expansion. Closer analogies between cable television and teletext are not as useful, however, because there are no other strong similarities. (One final note: as popular as cable television is to talk about, still less than one-third of all households subscribe to a cable service.)

Finally, a parallel for teletext development might come from the world of computerized information systems. This case is considered because of the possibility that, lacking readily available low-cost teletext–television sets, teletext will develop principally as an information service for business and industry. Historically, the use of computer terminals to access information stored in remote computers began to grow in the 1960s as computer networks began to be established. In the 1970s, the "data base industry" began to be recognized as such, with hundreds of sources of computerized information available to clients, largely in business, who had a computer terminal and a telephone. During the latter part of the 1970s, several computer services began to address the home market because personal microcomputers could be used as the terminal with the addition of a few pieces of equipment. After a few years, however, it appeared that the real market for large data base and data processing services was still the business world, because businesses are willing to cover the cost of accessing and possibly processing information stored remotely because there may be real profits in doing so. The home consumer sees a less direct benefit and often considers the cost of a computerized information service as a hobby expense (entertainment or self-education).

If the cost of the receiver remains above a nominal level as teletext develops, then teletext should look to the example of the computerized information services. The teletext technology can be used primarily for commercial purposes (e.g., to deliver information to clients who desire to receive and pay for it). Thus the service is based on the quality and desirability of the specific information being provided. The transmission mechanism is almost inconsequential, except that it may be less expensive or easier to use than a connection requiring a telephone to be dialed to call the remote computer.

There are many parallels, then, that we might find between teletext and past technologies. There are no absolute parallels certainly, because social, political, and economic conditions differ—not to mention the particular features of each technology. But some general lessons can be drawn. The link between the goals of a television network and the goals of a related manufacturer of television receivers will be the same in the case of teletext. The uncertain nature of the advertising potential of teletext suggests that teletext growth could have a rougher start than color television. The possibilities for teletext as a transmission and distribution facility for computerized business information services are there if the mass market applications are initially hindered. Teletext is unlikely to be as popular so quickly as radio because teletext lacks the major advantages that underlay the introduction of radio. Delays in the introduction of teletext whether because of war or lack of standards are unlikely to have any detrimental effect in the long term. As Table 7.1 indicates, the expectation of a ten-to-twenty-year period between the beginning of growth and widespread use seems reasonable.

The past, instructive as it can be, should be weighed in combination with an assessment of the present and near future relevant influences. Teletext is beginning to grow during a time when technology offers many alternatives and competing systems. The following section looks at some of these.

New Improved Television

The advent of teletext is part of an entire spectrum of technical advances that are just beginning to affect television. Putting teletext into perspective, we note a number of products, such as video cassettes and videodiscs, that are beginning to penetrate the market and are much more attractive to consumers than a teletext receiver. And there are products, such as high-definition television, that will enter the market in the near future but are already exciting interest among the more technology-oriented industry observers.

High-definition television uses techniques for creating television pictures with many more scan lines. While in Europe 625 scan lines per frame are used, and in North America 525 scan lines are used, the new high-definition television may have 1,125 scan lines. In addition to the use of about twice as many scan lines, the ratio of the horizontal and vertical dimensions of the screen may also change. The high-definition television system designed largely by NHK and several other Japanese television organizations has a horizontal/vertical ratio of 5 to 3 (the ratio of television today is 4 to 3). High-definition television also comes with enhanced color quality and stereo audio, and

Table 7.1. Growth Patterns for Several New Technologies (significant dates listed)

Year	Radio	Television	Color TV	Cable TV
1910	1910	1910		
	1916			
1920	1920			
	1922			
1930			1930	
1940		1940		
		1948		
1950			1950	1950
		1952		
			1953	
			1956	
1960				
			1966	
				1970
1970				
				1975
1980				1980

. . . Preliminary development.
---Growth begins.
⟶Widespread acceptance apparent.

it might use a video bandwidth as wide as 30 MHz. Much of the development of high-definition television has been done by Japanese companies—Sony, Panasonic, Ikegami, NHK—which have been working on the problem since at least 1968.

The excitement that surrounds high-definition television stems not only from viewers' anticipation of better quality television pictures, but also from producers' beliefs that the high-definition system will change "film making" considerably. The new system will permit the flexibility in image manipulation and editing that is possible with digitally stored pictures to be joined with the color and image quality that is endemic to film. Director Francis Ford Coppola has even gone further in his support for high-definition television, suggesting that the new television may be the most creative tool yet, because it permits the continual recombining of information (the electronic representations of images) that is the essence of creativity [9].

The relationship between high-definition television and the teletext technique exists in at least two ways, although in reality few people if any currently speak of both teletext and high-definition television in the same conversation. On the first level, because high-definition television is still television, the use of the teletext technique is still appropriate, and a successful application of teletext with today's television may be equally successful with tomorrow's television. But on another level, high-definition television can have a fundamental effect on teletext because the new television can provide the means for putting many more words on a single screen using smaller but easier-to-read characters. The present limitation of about 800 characters, or roughly 120 words, per screen could be overcome. With present teletext, the amount of information on a screen is less than that on one-half of a typed, double-spaced, standard letter-size paper; but with high-definition television, the amount of information might be five times as much, because high-definition television presents five times as much picture information as NTSC 525–line television. Perhaps the enhanced color quality will allow even more text per screen.

In the area of standards, the development of high-definition television gives the world another chance to agree upon a single, universal standard for television. If that happens, it is possible that a concurrent teletext standard could be agreed upon, although the likelihood of the latter is less than the probability of a single high-definition standard.

The future for high-definition television is as uncertain as that of any other new technology marked by high interest and high costs in the early years. One study, though, by Kalba Bowen Associates, projects about 30 million high-definition television sets in use by the year 2000 [10]. And the cost of a high-definition television receiver is predicted to start at $2,400 per set in 1984, diminish to $1,800 in 1985, and drop to $500 per set in 1999. These cost projections are not too much higher than the costs for a teletext-equipped television set. The 1982 cost of a Zenith teletext–television (which is not in volume production) is about $1,700. In a few years, if a consumer intends to buy a new television set and faces the choice of one set that can receive teletext and a second set at $500 more that can receive high-definition television (assuming that the two technologies have not been joined in the same piece of equipment), there is probably a better chance that the consumer will choose the high-definition set, on the basis that the choice is an entertainment decision and the extra money is not critical. But it could also be possible that if teletext circuitry cost is minimal, the decoder/character generator will be added to all new television sets whether conventional or high-definition television. However, the fact that black and white television sets are still manufactured and sold, as are sets without remote control and the like, suggests that that will not be the case.

The subject of high-definition television is intimately tied to several of the other topics affecting teletext, such as direct satellite broadcasting, cable television, video cassettes, and videodiscs. The Kalba Bowen study mentioned above suggests that high-definition television will first be delivered to television sets via cable and through the use of cassettes and videodiscs before the broadcasting of high-definition signals begins. When such broadcasting does begin, it is likely to be through the use of direct broadcast satellites, where the extra bandwidth is available because a channel of 75 MHz may be needed (which is over ten times the bandwidth of normal television).

Aside from the competition for the consumer's money among these new technologies, there is a good chance that some of these technologies can provide exactly the same service as teletext. For example, videodiscs (optical videodiscs specifically) have been considered for several years as a means of accomplishing digital information dissemination and retrieval for fairly large data bases [11]. (Optical videodiscs are phonograph recordlike discs with highly reflective surfaces. The surfaces are coded when microscopic dots, or "pits," are impressed into the surfaces, which are then coated with a protective layer of clear plastic. This coding technique is more than a simple binary digit scheme because both the length of the pit and the distance between pits is measured and used to convey information. The discs are "read" by measuring the reflections from a laser light focused on the disc's surfaces.) The amount of information that can be recorded on a videodisc approaches 2 billion to 3 billion characters per side. This is equivalent to about 400 million words. An optical videodisc thus encoded can be placed in a player under control of a microcomputer, giving the user the ability to search interactively for words or subjects contained within the disc's data base.

The immediate advantage of using optical discs as a means of information dissemination/retrieval is that it could be much less expensive to place periodically a computerized data base on disc and mail the discs to clients than to require clients to telephone a remote computer every time a piece of information is sought. In the telephone-computer system, the user commonly pays for the telecommunications link to the computer and pays a running connect-time fee to cover the cost of a computer system large enough to handle the interaction with all the remote users and all the interaction with the data base. The costs to the user can rise dramatically as more and more time is spent searching the data base. In the distributed disc system, the user does not pay for a telecommunications link and does not pay a connect-time fee that continually accumulates, because all interaction with the data base is done locally on the user's own disc player/microcomputer. Moreover, the central computer complex does not then need to handle a large number of incoming telephone calls or engage in searching the data base except for changes, additions, and deletions in time for the next "printing" of the discs to be sent out.

Although the potential of optical videodiscs as a replacement for current on-line computerized information systems is fascinating because of rather fundamental changes in the information industry that could eventually emerge, such speculation is beyond the scope of this book. However, there is a direct connection to use of the teletext technique that can at least be identified. One of the possible uses for teletext is as a means for disseminating digital information, and more particularly, disseminating the data to microcomputers. In much the same way that videodiscs can be used to bring a large computerized data base to a local data base-searching machine (the microcomputer), teletext can be used to transmit entire data bases to microcomputers with large memory devices.

For any information dissemination application, where the data base changes only gradually (i.e., not an up-to-the-minute news service), the user merely needs to receive the data base once a day, or once a week, or once a month, and so on. Assuming that four scan lines in the vertical blanking interval are available for teletext, over 21 million characters can be transmitted per hour. If an entire day is used to transmit a data base, one-half billion characters can be sent. In the cable television environment where a full television channel can be used for carrying teletext on each of approximately 200 scan lines, the transmission rate just mentioned can be multiplied by 25. Obviously, teletext can compete with the videodisc as the distribution medium for data bases as long as the price is competitive. On the surface, teletext would seem to have an advantage in that the distribution medium is not a physical unit as is the disc; in the disc distribution system, quantities of discs must be manufactured, encoded, distributed, and probably returned for reuse. The teletext system would not have such problems.

Other new forms of television that could affect, either positively or negatively, the future of teletext revolve around distribution techniques for television, such as low-power television, direct broadcast satellites, and multipoint microwave distribution systems commonly known as MDS but also called "wireless cable," among others.

Low-power television, mentioned in previous chapters, is just beginning to be authorized in the United States. If the number of low-power television stations reaches the predicted amount, there could be another 5,000 broadcast television stations across the country, in addition to the existing approximately 1,000 stations. This, of course, greatly expands the market for broadcast teletext systems if any significant proportion of the low-power stations consider adding teletext to their broadcast fare. But another aspect of low-power television is that many stations are likely to be part of one or more networks for programming and, perhaps, for teletext. This could affect the type of teletext service provided. If, on the other hand, low-power stations themselves become largely locally oriented outlets, the related teletext service may be much more local (e.g., the equivalent of the neighborhood news sheet or the supermarket bulletin board).

Low-power television also holds the promise of full-channel broadcast teletext. Either on a part-time or full-time basis, the bulk of the bandwidth of the broadcast channel could be used for transmitting digital information. For business customers, this could be the vehicle for distributing large data bases as described above. Businesses that already subscribe to computerized information systems could be supplied with the same service delivered in a different way. The fact that the data are broadcast (i.e., theoretically radiating to everybody within a certain geographic boundary) does not necessarily hinder specialized services because the pages of data can be addressed to individuals and to groups, and they can be encrypted if necessary. A low-power television station, using full-channel teletext, could broadcast medical information to doctors, stock and bond information to financial organizations, and so on, within the same television signal. The Federal Express Corporation, for example, has applied for low-power television stations for subscription teletext during the day, pay TV in the evening, and facsimile transmission in the early morning hours.

Direct broadcast satellite (DBS) systems beam television signals by satellites to television sets themselves (as long as the set has the proper antenna). Under present procedures in the United States, if a television signal is transmitted via satellite, it is normally received by a large dish antenna at a cable television headend or at a satellite receiving station, to be redirected either down a cable or over the air through a local television station's own transmitting tower. However, the idea of devising a broadcast television system such that the receiving antenna could be small enough to fit on a rooftop and yet the signal would be coming from a satellite in orbit 22,300 miles high has been discussed and tested over the past decade. Direct satellite broadcasting is being developed for home use in Japan, Canada, England, France, and West Germany and has been tested in such areas as remote parts of India and other developing countries. In the United States, after some experimentation during the 1970s, final rules for direct broadcast satellite service were approved by the FCC in 1982. At the same time, however, direct home reception of satellite television has been occurring on an unofficial basis for several years, as companies have offered for sale, and consumers have purchased, backyard-sized dish antennas. The estimated number of such home satellite receivers ranges from 10,000 to 40,000, and the price has dropped to less than $10,000. On the broadcasting side, at least nine companies have asked the FCC for permission to begin a DBS service, including television companies like CBS and RCA, as well as telecommunications companies such as Western Union and Satellite Television Corporation, a subsidiary of Comsat.

The advantage of direct television broadcasting via satellite is that many people who are not now served by either cable television or broadcast television will be able to receive the television signal beamed by a satellite. An estimated 6 million households in the United States do not have adequate television service (i.e., some cannot receive any television at all and the rest receive only one or two channels). DBS service is yet another means of expanding the number of channels available to the consumer, whether in a poor television service area or not.

Another use of DBS, however, is to transmit the new high-definition television, as previously mentioned. The Kalba Bowen study of high-definition television predicts that 20,000 households will receive high-definition television via DBS in 1987 (the DBS service itself will not be in operation before 1984–1985) and 160,000 will receive it by 1990 [12].

Again, the effects of DBS are in the not so near future, and the specific effect on teletext is uncertain. Teletext could be a part of DBS, expanding the capability of the relatively few DBS channels. Or DBS could have a negative effect on teletext in that DBS is a competitor for development resources, both in terms of an organization's technical expertise to devote to new projects and in terms of the organization's financial ability to do so.

Yet another technology for increasing the total number of television channels available in a given location is expansion of the multipoint distribution service (MDS). The Microband Corporation, a subsidiary of Tymshare, Inc., has requested permission from the FCC to create five new MDS channels in each major television market, and to link the markets together via Tymshare's national data network in a system to be called Urbanet [13]. The primary intent in creating the new MDS stations is to offer premium programming, or pay TV, in much the same way that cable systems do. But the five channels would also be used for data distribution using the teletext technique.

Conceptually, the Urbanet stations fall somewhere between broadcast television and cable television. On the one hand, a channel is broadcast as a single entity; on the other hand, a collection of channels is presented as a single multilevel service. The Urbanet concept is that the five channels are to be marketed as a single multilevel service. In the same manner, the organization of the teletext portion could lie somewhere between the cable example and the broadcast example. While the teletext service might lack the feedback feature of a cable system (unless return signals are routed via telephone lines), the vertical blanking intervals of the five channels could be treated as a single set for text transmission, or one of the channels could be used for full-channel teletext. Even if only the vertical blanking intervals are used, Urbanet could broadcast five times as many text pages in the same maximum cycle time as a single broadcast television station.

There are, in short, a number of new things happening to television as we know it, and these are summarized in Table 7.2. Some of these new features are inherently more exciting or more dramatic than teletext has been thus far. Probably all, or at least most, of the new developments in television could help or hinder the progress of teletext. However, it is still too early to see any of the specific trends.

In addition to these techniques and systems that are essentially apart from teletext, there are some as yet undeveloped aspects of the teletext technique itself that could also affect teletext's future. One such example is the addition of digital audio to the teletext data stream. As teletext is used today, there is no sound when a viewer switches to the teletext mode. (A possible exception is services such as Nite Owl in Chicago, where a music background accompanies the display of teletext-created pages, but this is not really teletext because the transmission method is normal television, not the teletext technique.) Work is already underway to transmit both digital data and digitized audio along the same telephone line, and the same concept could apply to teletext data in the vertical blanking interval.

But the audio feature does not necessarily require complete audio signals to be fully digitized and added to the existing data stream. After all, the vertical blanking interval does not have enough space for a great amount of data, at least not for the amount of data necessary for digitized audio. Nevertheless, audio features can be added in other ways. John Lopinto of Time, Inc., suggests that control signals could be inserted into the teletext data stream that could activate devices built into the teletext–television sets [14]. On a relatively simple level, the control signals could activate circuits to produce the beeps, crashes, whirs, and other noises associated with video games. On a more complex level, the audio codes could activate speech-generating chips integrated into the teletext receiver.

Another addition to the teletext technique is expansion by use of the audio bandwidth that is part of the television signal. The SCA services, discussed in previous chapters, use a portion of an FM radio signal to carry additional audio and digital data. Because television uses FM audio, the already available SCA procedures can be used to add either analog audio or digital pulses to the FM portion of a television signal. Alternatively, other similar procedures could be used such as that tested by Oak Communications to put teletext data onto an audio subcarrier as part of an STV broadcast channel.

In short, teletext is just beginning to grow within an environment full of other new television-related technologies that are certainly rivals for the public's attention, if not competitors for the same resources. And yet while the near future for teletext is conse-

Table 7.2. Some Related Television Technologies

High-definition television
Optical videodiscs
Video cassettes
Low-power television
Direct broadcast satellites (DBS)
Multipoint distribution service (multichannel arrangements)

quently uncertain, the long-term future for the teletext technique is nonetheless worth considering.

Social Impact

It is often difficult to assess the importance of contemporary events unless we look far to the past and then far to the future to see where we fit within the scale. Carl Sagan has produced a fascinating one-year calendar encompassing the entire 15-billion-year history of the universe; and the origin of humans does not take place until approximately 10:30 P.M. on December 31 of that year [15]. Although it is not helpful to take such an extreme yardstick to such a relatively trifling matter as teletext, it may indeed be useful at least to consider teletext within the context of the history of human communication.

Television itself, as any reader of Marshall McLuhan or other similar writers will know, has been awarded a pivotal position in the history of communication. Television, says McLuhan, is causing a retribalization of society, bringing back the information exchange patterns of tribal societies [16]. Speech communication is a low-definition process because it requires high participation by individuals. Print, however, along with radio, movies, and photographs, is a high-definition medium requiring rather low participation. McLuhan argues that the effects of the use of an alphabet, and later printing—which itself has been related to significant social changes—can be understood by recognizing the requirements of the medium (i.e., either high definition with low participation (hot media) or low definition with high participation (cool media)). He maintains that television, along with other media such as the telephone and cartoons, is a high-participation medium, and therefore the resulting communication patterns or social activity related to the use of the medium will ultimately be more like tribal society than modern society since the introduction of printing.

Putting aside for the moment any debate over the details of McLuhan's thesis, and assuming that there may be some truth in his overall characterization of communications media, the role for teletext is ambiguous. Because teletext is a television display it could share the characteristics of a low-definition (high-participation) medium, and yet because teletext is primarily textual, it could share the characteristics of a high-definition (low-participation) medium. On the basis that teletext is a subset of television, this implies that McLuhan's predicted social impact for television will be less true as television begins to migrate from one classification to a midpoint between the two. (This is based on an assumption that teletext will eventually be widely available and widely used.) Eventually, therefore, the major social effect of television may be altered by teletext, as television itself changes from primarily an image medium to both an image and text medium.

More immediately, teletext could spark a slow change in the information gathering habits of the general population. It has often been said, as television has grown, that television viewing could ruin reading. This has not been proven to be true, however. Books and magazines and other print publications continue to be sold and to be read in ever increasing numbers. In specific instances, of course, there have been declines in numbers of items published, and in one case there may be somewhat of a causal link between television viewing and reduced reading, namely, in the case of the evening newspaper. The health of evening newspapers in this country is declining, and at least one contributing factor may be the success of evening television news in combination with a general desire by people to relax after a day's work. While total newspaper circulation has remained fairly stable in recent times, there is a definite shift away from evening newspapers in favor of morning newspapers [17].

The substitution of one information source for another is not necessarily a bad thing, but critics of television news point out that fewer stories, in much less depth, are presented on television than appear in the evening newspaper. The substitution is thus unequal. Now, however, there is the possibility that teletext could change that. Teletext could provide the text in depth to support both the news items given video coverage and other items mentioned only by category. For example, during the evening news, the newscaster may refer to further details of a story on page 123 of the teletext magazine, or to more international stories from the Far East and Australia, for instance, on pages 12–34. The viewer can elect to read the items when something else comes on the screen that is less interesting, perhaps local sports, or to read the items later in the evening. The news show itself provides a headline service and brief coverage of the most important items. The person at home, who reads a newspaper by glancing at headlines and first paragraphs and skipping over stories of no interest, can acquire just as much information by watching/reading the evening television/teletext news as by reading a newspaper. (Teletext would have a more difficult time displacing the morning newspaper because, with current technology, one cannot take the teletext set on the bus or train for the commute to work.)

Another way to look at the larger implications of teletext is to consider its role within the history of creativity and social progress. The philosopher Alfred North Whitehead, among others, has explained persuasively how a certain amount of instability, a certain amount of difference and danger, is a prerequisite for the discontinuous social and technological progress that marks human history [18]. This same belief has been taken by Eric Somers, expressed as cultural energy, and related directly to the use of microcomputers, possibly receiving data via teletext [19].

Somers's thesis, built on the writings of Marshall McLuhan and others, is that television communication, in being similar to the communication patterns of tribal society, leads to a reduction in cultural energy. As people spend more and more time watching television, they become increasingly "normalized" (i.e., they reflect the average ideas and values that are presented on television as part of the broadcasters' desire to be attractive to the average). Because television broadcasting in the United States is ruled by audience ratings and the need to win the greatest audience share, which in turn is needed to attract advertisers who will support the programming, the overall effect is a reinforcing of any inclination to abandon social or cultural differences. And this, says Somers, has reduced the cultural energy that feeds creativity and social progress.

However, the recent proliferation of personal computers—home microcomputers and microprocessors—can change all that, according to Somers. The personal computer can be used to process and manipulate information or data, and this act of recombination and looking at things differently increases cultural energy. This occurs because the user is no longer accepting information as received but values even more the ability to change, to process, to recombine, and to look at the information in other ways, influenced by pieces of information acquired from other sources. In addition to the processing power of the personal computer, there is (or can be) a complementary capacity for storing massive amounts of digital information in readily accessible forms (e.g., on computer discs or videodiscs).

Although Somers notes that the use of microcomputers does not yet even approach the potential he has outlined, he suggests that it *could* happen if the designers and manufacturers of equipment and systems consciously try to create products that would increase cultural energy. The principles they would adhere to would be that the personal computer should have sufficient processing capability to manipulate large data bases, that the personal computer should be able to store large data bases, and that the personal computer should be able to receive information via many media.

The belief in an immense potential for microcomputers is based, of course, not on what microcomputers are today but on a more general idea of personal computing power. Carl Sagan suggests that the personal computer could possibly play a very important role in the overall progress of civilization. In fact, the possible effects of the proliferation of personal computers are equated with all the changes in society wrought by the invention of printing [20].

The relationship of all this to teletext is that teletext could provide the low-cost distribution facility for transmitting information to the personal computer. Doubtlessly by the time personal computers begin to affect society in the ways suggested, the teletext technique will have long since disappeared as such in the continuing shift to the digitizing of all forms of electro-optical communications, and to increased utilization of all channels or media. But in the near future, the possibility exists for teletext to rival the telephone network as a means of disseminating information from central data bases or collection points to microcomputers/receivers. There is nothing very dramatic in using teletext in such a way; in fact, this is already happening to a limited extent. Broadcast teletext has been used for transmitting ''telesoftware'' (i.e., transmitting short computer programs to home microcomputers). In the cable environment, the teletext technique is used to disseminate business information to terminals (not necessarily microcomputers). But in the next decade or two, as traditional television channels increase in number, the potential for a rather large-scale application of teletext for information dissemination to ''intelligent'' or semiintelligent machines exists. Broadcast teletext has an advantage over other distribution media such as the telephone network by the very fact that it is broadcast. Microcomputer-controlled receivers can elect when to store the broadcast data and when not to, based on codes in the teletext data stream and on preset instructions in the receiver. The result can be a very dynamic information dissemination process. If a similar sort of procedure is established using the telephone network, either the line between the central computer and the microcomputer would have to be kept open (in use) all the time, or the receiver would have to dial the computer periodically. But if the receiver is set to dial at specific times, it cannot react to any changes in timing that occur at the central computer. However, the microcomputer picking and choosing from broadcast teletext can do so.

Along with the relationship to microcomputers, or indeed to any of the pieces of equipment used to receive the teletext signal, teletext itself can be considered as part of the computer/telecommunications union that is behind the phrases information age, information society, information economy, postindustrial society, and the like [21].

Beginning with the invention of printing—the ability to reproduce text relatively easily—we have witnessed a gradual accumulation of storable information. With the perfection of both digital and nondigital techniques, the amount of stored information has increased tremendously during the past fifty to one hundred years. But more than that, during recent times there has been an increasing percentage of the work force that is primarily involved with information gathering, storing, processing, and disseminating [22]. According to some calculations, over half of all employees in the United States work in the information sector, about equally divided between workers directly associated with the information sector and those secondarily associated. In other countries, the size of the information work force is placed at 30 percent for Japan (in 1975), 34 percent for West Germany (in 1976), and 40 percent for Canada (in 1971) [23].

The implication behind this recorded growth in the information sector of our economy is that dealing with information is not the same as dealing with conventional products and services, at least not in economic terms. Many books and articles have been written about the fundamental changes in society that will eventually be attributed to the information orientation that has occurred in recent times and is still growing. And the tool, or facilitating agent, of the information exploitation has been the combination of computers and telecommunications.

Teletext, as a technique that combines computer power with telecommunications, is therefore a definite part of the information age phenomenon, however minute the use of teletext might be at present. Beyond that, teletext has intriguing possibilities because it joins together the characteristics of broadcast communications with point-to-point communications, and it also brings together the characteristics of print media with the characteristics of video media.

The very, very long-term effects of teletext as part of the computer/telecommunications union have been suggested above. Even further suggestions about an actual physical link between the human brain and "intelligent" machines could be postulated in the manner of Carl Sagan [24], Isaac Asimov [25], or Teilhard de Chardin [26]. But that would draw attention away from the fact that a very real change could be taking place in our own society over the next 5 to 25 years, based in part on the application of the teletext technique. Although problems of definition abound when concepts such as information, knowledge, energy, and the like are seemingly equated, there is a realization that computer/telecommunications systems, of which teletext is a part, are subtly and not so subtly shifting patterns of work and play.

Conclusion

Throughout this book there have been descriptions of what teletext is and how the teletext technique is used, and suggestions of what teletext might become. The term itself is not very old, perhaps eight or ten years old (unless etymologists can find earlier occurrences of the word). And in the early 1970s teletext was often used as a general term for any system for putting text on television screens, while now the term refers more specifically to systems for multiplexing digital information onto a television signal. The utilization of teletext as a vehicle for services is just beginning, led by the early work of the BBC and the IBA in Great Britain, and PBS in the United States.

Naturally, many people wonder not so much about the long-term future for teletext but about the very immediate future—the next year or two. Who will pay for teletext to get started in the United States? Or to put it another way, what aspects of teletext will be successful initially? What services should the entrepreneur concentrate on?

Walter Ciciora suggests that broadcast teletext will be successful if it provides entertainment and convenience, but he also points out the possibility that teletext should initially concentrate on special interest groups, such as captioning for the deaf [27]. Other possible special interest categories are sports information, summaries of the shows on the video portion of the television channel, and historical summaries of serial programs with story lines continuing for months or even years. Other observers have remarked on the intriguing implications of all kinds of program-related teletext (i.e., the video program refers to specific items in the teletext magazine, or the teletext pages must be viewed in conjunction with the video programming). [28]. During a football game, for example, viewers could be alerted to the presence of pages of background data contained in the teletext magazine, which could be viewed superimposed over the live video if the viewer so chooses. But the link between the video program and the teletext magazine can also be used for advertising; commercials could refer to additional text information in the teletext magazine, perhaps giving the viewer some incentive to look at the teletext pages (e.g., a discount of some sort).

The advertising potential of teletext is one of the biggest unknowns. There are seemingly good reasons for both sides of the argument (i.e., that advertising will be successful on teletext, and that it will not be). In support of the latter side, critics note that the low-cost teletext systems have elementary graphics capability, such that advertisers cannot rely on a good graphic presentation to attract attention and potential buyers. And teletext systems with good graphics capability just might be too expensive to generate a mass audience quickly. On the other hand, as teletext grows beyond the very first stages, advertisers will know who the audience is (as was the case with color television for about ten years)—the audience is in the upper-income brackets with an interest in new devices, a certain amount of discretionary income, likely to purchase certain categories of products, and so on. Thus advertisers seeking such a market could shift their money to television stations broadcasting teletext, which will in turn promote the use of teletext and the purchase of teletext sets in order to strengthen their own market position.

Advertising, though, is only one potential source of revenue for teletext. As has been discussed, the teletext technique can be used for virtually any form of information dissemination, and existing information services could shift to teletext if the overall costs will be less. There is ample support from the history of similar systems for the suggestion that teletext will experience several years of growth primarily in the special services area. In other words, the teletext technique will be successfully used as a means of distributing information for relatively small groups of users, in situations where there is an identifiable audience with a real need and an ability to pay.

Aside from the advertising and information service revenue, there is always money to be gained from games and entertainment applications. Teletext could be used to transmit the digital information to create the type of home video games that we are familiar with in cartridge form. In the program-related sense mentioned above, teletext could be used to enhance game shows in a variety of ways, from providing more information of which the on-screen game player is unaware to providing tasks that the home viewer can complete in order, perhaps, to win prizes.

Over the next five to ten years, though, the future for teletext is not expected to be spectacular. Several projections for the year 1990 place the number of teletext–television sets at 8 percent of the total or fewer. RCA's Consumer Electronics Division predicts that only 8 percent of U.S. households will have either teletext or videotex receivers by 1990 [29]. This is based on the assumption that the FCC will not promulgate a definite teletext standard (at least not soon), and that no regulation will be enacted to require the addition of teletext circuitry to new television sets. Another forecast, by CSP International, indicates that while 10 percent of all color television sets sold in 1990 will have teletext capability, this would represent fewer than 4 percent of all television households [30]. Even in England, where the television organizations and the television manufacturers agreed upon a standard, and over 60 percent of the population rent rather than buy their television sets (thus making the acquisition of a teletext-equipped television set much easier), the number of teletext households is only about 3 percent of the total television households seven years after teletext broadcasting began.

It is quite likely that broadcast teletext will be dominated by the major television networks. This could be inferred from the very nature of teletext as a technique for utilizing television signals, but the assumption also gains force from the fact that the three commercial television networks top the list of the largest media companies in the United States, based on media revenue [31]. The three networks are followed on the list by Time, Inc., which has begun a teletext-for-cable national service, and by eight companies that are primarily newspaper publishers. If the large media companies are indeed the ones that will most affect the near-term future for teletext, there may be some chance that the newspapers will have almost as much influence as the television networks, given the textual nature of teletext. The interplay would be between the networks that dominate the medium and the newspaper publishers who control the content (here the assumption is that newspaperlike information will be the staple of broadcast teletext). If the two groups have different goals, the greater weight will probably be on the side of the medium owners—the networks and their affiliated stations.

This, then, is the beginning of the story. The groundwork for teletext has been done. The potential uses for the teletext technique have been demonstrated. Teletext services are being established. In the large view, teletext is part of the computer/telecommunications force that is setting society on a new tack. On the less dramatic scale, teletext is a relatively new addition to the television experience, and only one among many. In the next several years, the appearance of television's teletext will be shaped by the decisions of a few and the reactions of the rest, as teletext services begin to grow.

References

1. Aronson, Sidney H., Bell's Electrical Toy: What's the Use. In *The Social Impact of the Telephone,* Ithiel de Sola Pool (ed.), MIT Press, Cambridge, 1977, pp. 19–20.
2. Emery, Walter B., *Broadcasting and the Government,* Michigan State University Press, East Lansing, 1971, p. 20.
3. Barnouw, Erik, *A Tower in Babel. A History of Broadcasting in the United States,* vol. I, Oxford University Press, New York, 1966, p. 79.
4. Ibid., p. 116.
5. Personal comment of Thomas O'Brien, Jerrold Division of General Instruments Corporation, 1981.
6. This observation has been made by, among others, a former BBC employee, Gwyn Morgan, who has written a number of articles about teletext.
7. A good summary of the history of broadcasting is contained in the Federal Communications Commission's "Broadcast Primer. Evolution of Broadcasting," Inf. Bulletin #2–B, July 1966.
8. These and following observations on color television are largely from Howard W. Coleman (ed.), *Color Television,* Hastings House, New York, 1968.
9. Rebuttal Witness, *Broadcasting,* February 1, 1982, p. 84.
10. Research Firm Sees Bright Future for HDTV, *Broadcasting,* February 1, 1982, p. 88.
11. For a good summary article, see Charles M. Goldstein, Optical Disk Technology and Information, *Science* 215: 862–868 (February 12, 1982).
12. Research Firm Sees Bright Future for HDTV, *Broadcasting,* February 1, 1982, p. 88.
13. Pollack, Andrew, Expanded Pay TV over the Air is Sought, *The New York Times,* February 11, 1982, p. 31.
14. Here Comes Talkie-Text, *Communications Daily* 1(210): 1–2 (November 16, 1981).

15. Sagan, Carl, *The Dragons of Eden,* Ballantine Books, New York, 1977, pp. 14–16.
16. See Marshall McLuhan, *The Gutenberg Galaxy,* University of Toronto Press, Toronto, 1962, and *Understanding Media,* McGraw-Hill, New York, 1964.
17. *Editor and Publisher 1981 International Yearbook,* Editor and Publisher, New York, 1981, p. v.
18. Whitehead, Alfred North, *Science and the Modern World,* Free Press, New York, 1925, pp. 206–207.
19. Somers, Eric, Cultural Energy and the Personal Computer, *Computer Age,* January 1980, pp. 8–12.
20. Sagan, p. 232.
21. See, for example, Richard H. Veith, *Multinational Computer Nets,* Lexington Books, Lexington, Mass., 1981, pp. xi–xvii.
22. See Fritz Machlup, *The Production and Distribution of Knowledge in the United States,* Princeton University Press, Princeton, N.J., 1962); and Marc Uri Porat, *The Information Economy: Definition and Measurement,* report prepared by the U.S. Department of Commerce, Office of Telecommunications, 1977.
23. Veith, p. xiii.
24. Sagan, p. 236.
25. Asimov, Isaac, Communication by Molecule, *Bell Telephone Magazine,* March–April 1973, pp. 7–11.
26. Teilhard de Chardin, Pierre, *The Phenomenon of Man,* Harper & Row, New York, 1959, p. 240.
27. Ciciora, Walter S., Teletext Systems: Considering the Prospective User, *SMPTE Journal* 89 (11): 846–849 (November 1980).
28. Arlen, Gary, The Revenue Potential of Teletext, *View,* June 1981, p. 91.
29. Research, *International Videotex Teletext News,* December 1981, p. 9.
30. CSP International, Market Analysis: Teletext in the United States, October 1981, p. 33.
31. Winski, Joseph M., Media Movers: Powers that Are, *Advertising Age,* December 7, 1981.

AI

Broadcast Specification for Canadian Teletext

Reprinted from the Government of Canada, Department of Communications, "Broadcast Specification: Television Broadcast Videotex," BS–14, Issue 1, Provisional, June 19, 1981. Used with the kind permission of the Canadian Department of Communications.

Television Broadcast Videotex

Definition

Television Broadcast Videotex: A system consisting of a central data store (data base) from which digital data representing text and pictorial information is transmitted in the active portion of available TV lines through a broadcast delivery system. User terminals then interpret and display the selected data on video receivers/monitors or other terminal devices.

Introduction

The parameters outlined in this document have been selected intending to comply with the following principles and requirements:

i. Terminal independence; this permits the use of a variety of terminals of varying capabilities, such as different levels of resolution.
ii. Compatibility between services carried over existing communications networks (e.g., public switched telephone, off-air broadcast, satellite, and cable TV networks) and common presentation format.
iii. Vertical blanking interval (VBI) and full field transmission compatibility.

iv. Forward and backward compatibility; permitting future terminals to access old data and requiring that an installed inventory of terminals be able to receive and decode all future command formats in an intelligent manner.
v. Adherence to already established national and international standards such as those contained in Appendix A.

Applicability

This document sets forth the requirements for the issuance of a Technical Construction and Operating Certificate (TC & OC) for a broadcasting transmitting undertaking when transmitting digitally encoded data for purposes including alphanumeric and/or pictorial information. The requirements also apply to a broadcasting receiving undertaking when the distributed signals are received from a broadcasting transmitting undertaking.

1.0 *Data Positioning and Waveforms*
Data may be transmitted in the active portion of a television line, commencing after the standard NTSC line synchronization and colour burst.

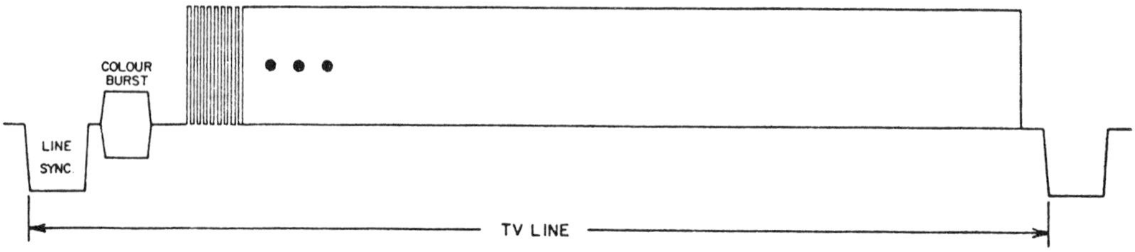

Figure 1

1.1 *Vertical Blanking Interval and Full-Field Data Transmission*
Data transmission uses the field-blanking interval and/or the active part of the video signal. Lines 1 through 21 in field 1 and the corresponding lines in field 2 of the 525 line 60 field/sec M/NTSC television system are designated as the vertical blanking interval (VBI). Of these, the allocation of lines 10 through 21 in field 1 and the corresponding lines in field 2 is the subject of Broadcast Specification 13. Full-field data transmission is achieved through utilization of lines 10 through 262 in field 1 and the corresponding lines in field 2 which comprises the vertical blanking interval as well as the active part.

1.2 *Transmission Bit Rate*
The transmission bit rate is 5,727,272 ± 16 bits/second[1] which is the 364th multiple of the horizontal line scanning rate for colour transmission (15,734.264 ± .044 Hz) and 8/5 of the colour subcarrier frequency (3,579,545 ± 10 Hz). The data signal is to be phase locked to the colour subcarrier when inserted into a colour television transmission and to the horizontal line scanning rate when inserted into a monochrome television transmission (with no burst present). The

[1]*Note:* The adoption of the proposed bit rate of 5,727,272 ± 16 bits/sec will be finalized only following a period of adequate experimentation. Should it be determined that the proposed bit rate is unsatisfactory in providing adequate service, an alternate bit rate would be considered. This action would therefore necessitate the revision of related parameters as presented in this provisional document.

maximum rate of change of the transmission bit frequency shall be 0.16 bits/second/second.

1.3 *Data Encoding*
The amplitude modulated data are nonreturn to zero (NRZ) binary encoded. Other encoding schemes are for further study.

1.4 *Data Pulse Shape*
The Spectrum of the NRZ data after shaping and Impulse Response of the Nyquist filter have the following characteristics:

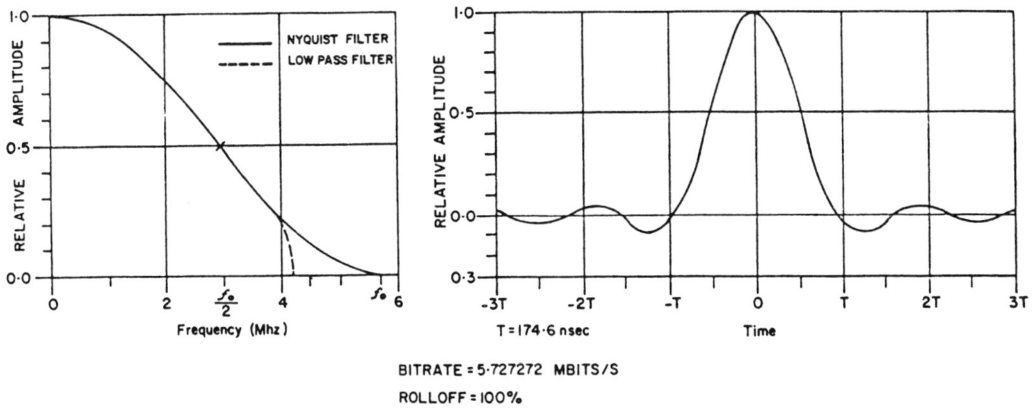

Figure 2

The data spectrum of the controlled raised cosine filter is described as follows:

$$H(f) = \begin{cases} 1 & \text{for } f < (1 - R)\, f_o/2 \\ \dfrac{1}{2}\left[1 - \sin\left(\dfrac{\pi}{2}\,\dfrac{f - f_o/2}{R\, f_o/2}\right)\right] & \text{for } |f - f_o/2| \leq R\, f_o/2 \\ 0 & \text{for } f > (1 + R) f_o/2 \end{cases}$$

The impulse response of the Nyquist filter is as follows:

$$h(t) = \frac{\sin \pi f_o t}{\pi t} \; \frac{\cos \pi f_o t R}{1 - (2 f_o t R)^2}$$

where

f = frequency in MHz
t = time in nanoseconds
f_o = bit rate in Mbits/sec
$f_o/2$ = center position of roll-off in MHz
R = roll-off = (i.e., 100%)

The spectral content of the shaped data is determined by a Nyquist filter with 100% roll-off, followed by a phase corrected low-pass filter with a cut-off of 4.2 MHz.

1.5 *Data Timing*

The half-amplitude point of the first data bit, as shown in Figure 3, is positioned 10.5 ± 0.34 μsec from the half-amplitude point of the leading edge of the horizontal sync. pulse.

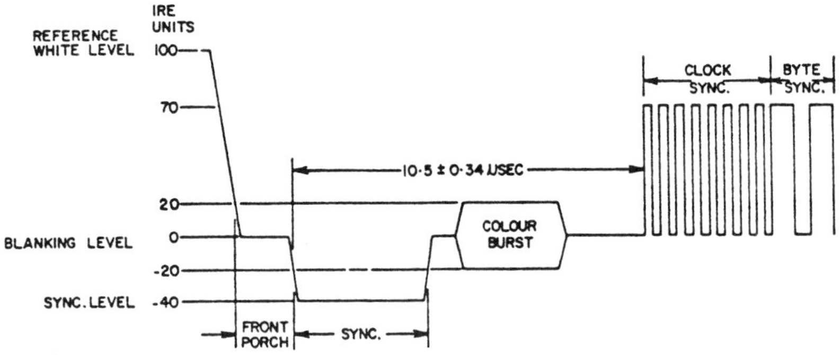

Figure 3

1.6 *Data Amplitude Modulation*

Data amplitude modulation parameters have been nominally established as 2 ± 2 IRE units for a logical "0" and 70 ± 2 IRE units for a logical "1," with provisions for positive and negative overshoots of 3 IRE units each. These nominal specifications permit a maximum peak-to-peak data amplitude of 78 IRE units.

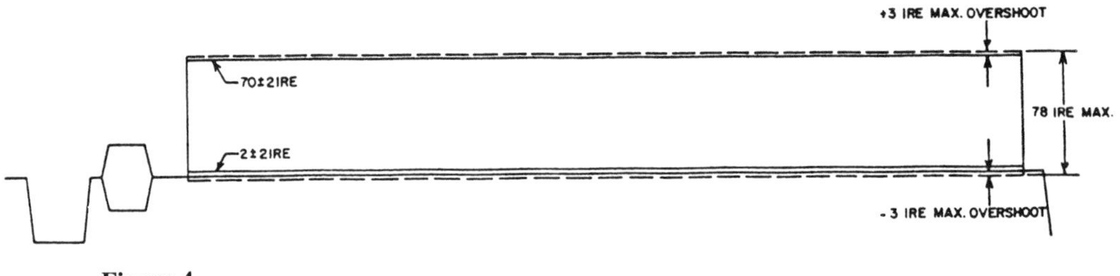

Figure 4

2.0 *Data Line*

The Data Line consists of a string of 288 bits (impulses) having the following format:

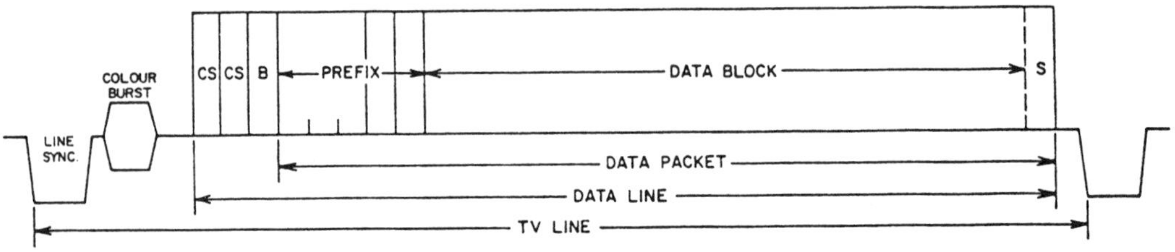

Figure 5

2.1 *Bit Synchronization*

The first 16 bits of the Data Line constitute the bit synchronization sequence (CS, CS) consisting of alternating 1's and 0's, leading in with "1." This sequence provides the decoder with a reference burst in order to synchronize the decoder's data clock and initialize the data slicer.

2.2 *Byte Synchronization*

The next 8 bits of the Data Line constitute the Framing Code (B) and serves to define the byte structure. This code has been chosen to minimize the potential of incorrect synchronization even in the presence of a single bit error in the Framing Code. The least significant bit (b1) is always transmitted first.

The sequence identified for Television Broadcast Videotex is:

$$\begin{matrix} 1 & 1 & 1 & 0 & 0 & 1 & 1 & 1 & \equiv 231_{10} \\ b_8 & b_7 & b_6 & b_5 & b_4 & b_3 & b_2 & b_1 & \end{matrix}$$

Two other compatible Framing Codes reserved for future use are:

$$\begin{matrix} 1 & 0 & 0 & 0 & 0 & 1 & 0 & 0 & \equiv 132_{10} \\ 0 & 0 & 1 & 0 & 1 & 1 & 0 & 1 & \equiv 45_{10} \\ b_8 & b_7 & b_6 & b_5 & b_4 & b_3 & b_2 & b_1 & \end{matrix}$$

3.0 *Data Packet*

The Data Packet is an identifiable package transmitted after the Bit and Byte Synchronization codes and is made up of a Prefix, a Data Block, and an optional Suffix.

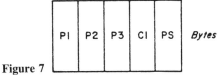

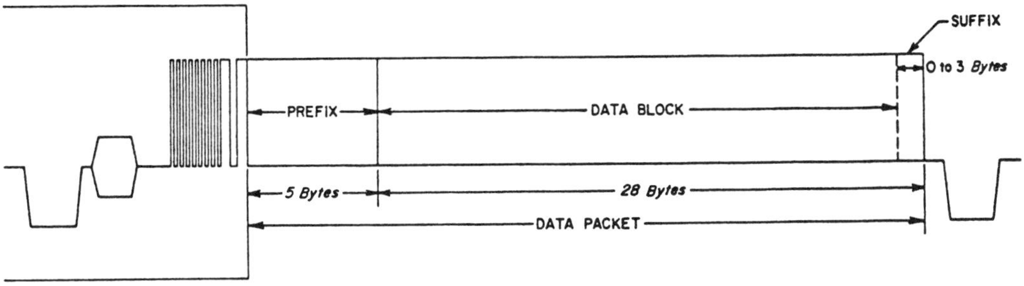

Figure 6

A Data Packet is contained within a single Data Line. Extended Packets, encompassing more than one Data Line, are for further study.

3.1 *Prefix*

The Prefix consists of 5 Hamming[2]-encoded bytes, the first 3 of which are Packet Address bytes followed by a Continuity Index byte (CI) and a Packet Structure byte (PS).

P1	P2	P3	CI	PS	*Bytes*

Figure 7

3.1.1 *Packet Address*

Three Hamming-encoded bytes (P1,P2,P3) constitute the Packet Address, yielding 2^{12} (4,096) possible data channels that may be time-division mul-

[2]In this document, an (8,4) Hamming data protection scheme is used, thus permitting single bit error correction, and even multiple bit error detection. In any Hamming-protected byte, bits b1, b3, b5, and b7 provide error protection and bits b2, b4, b6, and b8 present the information to be conveyed. Appendix B provides the Hamming Code Table.

tiplexed onto a single television channel. Channel numbers correspond directly to Decimal Binary Coding permitting user selection of the first 1,000 channels using a simple decimal keypad. The remaining channels are reserved for future use.

The scheme allows for interleaving of combinations of up to four (4) consecutive data lines, with one Packet Address, with Data Lines of other Packet Addresses and requires a minimum time separation of 4 milliseconds for both VBI and full-field operation.

3.1.2 *Continuity Index*

The Continuity Index consists of one Hamming-protected byte (CI) used to detect the loss of a Data Packet due to transmission errors. The Continuity Index sequences from 0 to 15 and is incremented by 1 for each transmission of a Data Packet within a Data Channel.

3.1.3 *Packet Structure*

The Packet Structure byte consists of a Hamming-protected byte (PS) specifying the nature of the transmitted Data Packet as follows:

Information Bits				Assigned Significance
b8	b6	b4	b2	
			0	Standard Packet
			1	Synchronizing Packet
		0		Packet full of information bytes
		1		Packet not full of information bytes[a]
0	0			No Suffix
0	1			1-Byte Suffix
1	0			2-Byte Suffix
1	1			3-Byte Suffix

[a]Bit b4 = 1 (Packet not full of information bytes) is not used to signal the end of the Data Group.

3.2 *Data Block*

The Data Block, which follows the Prefix, contains the Control Data (Header) and/or Presentation Data delivered to a terminal.

The Data Block may be full, or not full of information bytes, as indicated by bit b4 of the Packet Structure byte (PS).

If the Data Block is designated "Not Full," the noninformation bytes may be assigned "don't care" values; however odd parity must be maintained to ensure correct interpretation when a one-byte Suffix is employed.

The Data Block is reduced by the number of Suffix bytes as specified by bits b8,b6 of the Packet Structure byte:

no suffix byte; the Data Block contains 28 bytes,
1 suffix byte; the Data Block contains 27 bytes,
2 suffix bytes; the Data Block contains 26 bytes,
3 suffix bytes; the Data Block contains 25 bytes.

3.2.1 *Control Data (Header)*

The Header consists of information bytes used in instructing the terminal in processing Presentation Data.

3.2.2 *Presentation Data*

The Presentation Data is comprised of the data to be processed by the user terminal.

3.3 *Suffix*

An optional Suffix may follow the Data Block as determined by bits b8,b6 of the Packet Structure byte. This Suffix may contain one or more redundancy bytes that may be used by the data receiver for either error detection or correction in the Data Block.

A single byte Suffix is comprised of a longitudinal odd parity check of all bytes in the Data Block, which themselves contain an odd parity check in the most significant bit (b8) of each byte. This information forms the basis of the product code used to correct any single bit error and detect all double errors in each byte.

Other error detection/correction schemes for double or triple byte Suffixes are reserved for future assignments.

4.0 *Data Group*

Data Blocks associated with information from the same source (i.e., common Packet Address bytes: P1, P2, P3) may be sequentially organized into identifiable groups known as Data Groups. In broadcast videotex, these Data Groups are limited in length to a maximum of two (2) kilobytes.

The beginning of a Data Group is identified by bit b2 = 1 of the Packet Structure byte (PS). Each Data Group is composed of a Data Group Header followed by a Record.

4.1 *Data Group Header*

The Data Group Header follows the Prefix and is composed of the following Hamming-protected bytes:

TG	C	R	S1	S2	F1	F2	N

4.1.1 *Data Group Type (TG)*

This byte specifies the applicable class of processing to be applied by the data receiver.

TG = 0 designates the method of transmission used for broadcast videotex service. All other Type assignments are reserved for future use.

4.1.2 *Data Group Continuity (C)*

This byte is used to verify the sequence of Data Groups of a common Type (TG) in a particular Data Channel (P1, P2, P3). This continuity counter sequences from 0 to 15 and is incremented by 1 for each subsequent transmission of a Data Group of this nature.

4.1.3 *Data Group Repetition (R)*

This byte specifies the number of retransmissions of a given Data Group. This byte is restricted to the range 0–15.

4.1.4 *Data Group Size (S1, S2)*

Bytes S1, S2 specify the number of Data Blocks in a Data Group. These bytes indicate values ranging from 0 to 255.

4.1.5 *Last Block Size (F1, F2)*

These information bytes indicate the number of bytes in the last Data Block of a Data Group.

4.1.6 *Data Group Routing (N)*

A single byte, under broadcaster control, which identifies the routing of a Data Group through a broadcast network. Values in the range 0–15 may be assigned to this byte to control such functions as passage through time zone delay centers. This byte is not intended for use by the decoder.

5.0 *Record*

The Record is essentially the same as the Data Group stripped off the Data Group Header and contains information pertinent to broadcast videotex service. Each Record is comprised of a series of up to 256 sequentially numbered Data Blocks. The format consists of a Record Header, containing Record protocol information, followed by Presentation Data.

5.1 *Record Header*

The Record Header immediately follows the Data Group Header and is of variable length as determined by the Record Header Designator (RHD) and addition of optional subgroups (refer to Appendix C).

All Record Header bytes are Hamming-protected and are organized as follows:

TR RHD	A1 A2 A3 A4 A5 A6 A7 A8 A9	L1 L2	$Y0_1 ... Y0_N$, $Y1_1, ... Y1_6, ... YN_1 ... YN_6$	E1 ES $E_1 .... E_N$

← *Mandatory* → | ← *Optional* →

5.1.1 *Record Type (TR)*

This byte characterizes the type of information transmitted within a Record and its associated structure. Values have been defined as follows:

Decimal Value	Record Type
0	Cyclic Broadcast (e.g., Broadcast videotex)
1	Noncyclic Broadcast (e.g., Captioning)
2–15	Reserved for Future Use

5.1.2 *Record Header Designator (RHD)*

The Record Header Designator byte (RHD) is comprised of 4 flags that, when set, indicate the presence of one or more following subgroups that are concatenated in the following order:

b2 : Address Extension
b4 : Record Linking
b6 : Complementary Record Classification
b8 : Header Extension Field

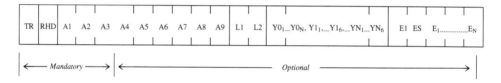

The status of each of the above subgroups is indicated by a single bit, where a binary "1" indicates the presence of the subgroup, and a binary "0" indicates its absence.

5.1.3 *Address Bytes (A1, A2, A3, A4, A5, A6, A7, A8, A9)*

Address bytes A1, A2, A3 represent a Record or Page number, and are considered mandatory. Address bytes A4, A5, A6, A7, A8, A9 are optional extension bytes that are transmitted when bit b2 of the Record Header Designator is set to "1." In this case, bytes A1–A7 are used to represent a Document number and bytes A8, A9 represent the number of a Page within a Document.

5.1.4 *Record Linking (L1, L2)*

The presence of the Record Linking bytes L1, L2 is indicated by bit b4 = 1 in the Record Header Designator. These bytes immediately follow the Address bytes. These bytes are used to link together Records identified by the same address and associated with the same message. The decoder must capture linked Records in sequential order.

Bits b6, b4 and b2 of byte L1 and b8, b6, b4, and b2 of byte L2 are used to indicate the order of the linked Records. Bit b8 of L1 is used to indicate the existence of additional linked Records (b8 = 1) or the last linked Record in the sequence (b8 = 0).

5.1.5 *Complementary Classification Sequence* ($Y0_1 - Y0_N$, $Y1_1 - Y1_6$, . . ., $YN_1 - YN_6$)

The presence of the Complementary Classification Sequence subgroup is indicated by bit b6 = 1 in the Record Header Designator. This subgroup follows the Address and Link bytes, if present, or the Record Header Designator for the case b2 = b4 = 0.

This subgroup has a variable format and is divided into two sections: refer to Appendix D.

- Designation bytes ($Y0_1$–$Y0_N$)
- Complementary Classification bytes ($Y1_1 - Y1_6$, . . . , $YN_1 - YN_6$)

Bit b8 of any Y0 byte indicates the status of any additional Designation bytes, where a binary "1" indicates the presence of an additional Y0 byte, and a binary "0" indicates the end of a Y0 sequence. For example, if bit b8 of $Y0_1$ is equal to "1" an additional Y0 byte designated $Y0_2$ exists. This pattern also applies to $Y0_2$, which in turn may indicate a $Y0_3$ byte, etc. The Y0 sequence is terminated only when bit b8 of the last Y0 byte is equal to "0." This scheme is illustrated in Appendix D.

The remaining usable bits (b6, b4, b2) of any Y0 byte point to a group of Classification bytes; refer to Appendix E. Each bit of a Y0 Designation byte is associated with a Classification field, either specifying a function by default ($b_x = 0$) or calling a function ($b_x = 1$) and thus specifying it with a two byte sequence.

When Y0 has bits b8 = b6 = b4 = b2 = 0, it is the end of the Record Header and all Record specification functions are specified by default.

5.1.6 *Header Extension Field (EI, ES)*

Any number of variable-length Header Extension Fields may be designated by b8 = 1 of the Record Header Designator (RHD) byte (see Appendix C). These fields directly follow the Address, Link and Complementary Classification bytes, if present, or the Record Header Designator for the case b6 = b4 = b2 = 0.

The first byte of the Header Extension Field is an extension field introducer byte (EI), with bit designations as follows:

b8 = 1	indicates further Header Extension Field(s) to follow (E_1 . . . E_N)
= 0	indicates last Extension Field
b6,b4,b2	indicate Header Extension Code assignments whose values are given in Appendix E.

The second byte (ES) of the Header Extension Field indicates the number of bytes of Extension Field information in the current Extension Field

5.2 *Segmentation*

The Data Record may be segmented by the use of a ''Segmentation Identification Sequence'' that consists of a three byte sequence; the first byte of which is a specific code corresponding to the 'US' (unit separator) code taken from the C0 code table at the presentation level. The remaining two bytes are designated for use by the presentation level and to indicate the relationship of segments.

6.0 *Repertoire*

The presentation coding scheme for text conforming to international recommendations from the C.C.I.T.T. and C.C.I.R. permits the coding of a large repertoire of characters and special symbols covering all Latin based alphabets. The languages of primary interest in North America are English, French, and Spanish. The character repertoire implemented in a particular terminal should contain the appropriate accented characters for these languages, as well as all of the characters in the CSA basic character set Z243.41 set 1 (known as ASCII in the USA). To fully present the French (for Canada) and the Spanish languages requires that all appropriate accents be displayed. Technical compatibility requires that all terminals correctly interpret all coded accents and special characters and provide at least the appropriate defaults for these languages. In addition, the following special symbols should be provided: << >> ¿ ¡ ɗ

6.1 *Presentation Coding*

The presentation coding scheme for Broadcast Videotex services is that adopted by the Canadian Standards Association (CSA). This coding scheme is based on the alphageometric coding scheme described in C.C.I.T.T. Recommendation S. 100 and the C.V.C.C. Videotex Field Trial Presentation Layer Standard No. 699 that are reflected in the proposed North American Standard Presentation level protocol.

6.2 *Display Format*

The default Display Format is defined as 20 rows of 40 alphanumeric characters per row within the S.M.P.T.E.[3] Safe Title Area of the television screen. Other display formats are also permitted.

6.3 *Display Attributes*

C.V.C.C. Videotex Field Trial Presentation Layer Standard No. 699 and C.C.I.T.T. Recommendation F.300 present the various degrees of implementation of the display attributes for videotex systems.

Issued under the Authority
of the Minister of Communications

Dr. John deMercado
Director General
Telecommunication Regulatory Service

Appendix I.A

The parameters outlined in BS–14 have been selected intending to comply with the following established national and international standards and the recognized principles and requirements contained therein.

International Telegraph and Telephone Consultative Committee (C.C.I.T.T.):

- Recommendation S.100 "International Information Exchange for Interactive Videotex"
- Recommendation F.300 "Videotex Service"

International Radio Consultative Committee (C.C.I.R.) Report 624–1 Characteristics of Television Systems (System M/NTSC)

International Organization for Standardization (I.S.O.):

- Draft International Standard ISO/DIS 2022 "Code Extension Techniques for Use with the ISO 7-bit Coded Character Set"
- Draft International Proposal ISO/DIP 6937 "Coded Character Set for Text Communication"
- ISO/TC 97/SC 16 N 537 "Basic Specifications of the Reference Model of Open System Interconnection"

Canadian Videotex Consultative Committee (C.V.C.C.) Videotex Field Trial Presentation Layer Standard (Communications Research Centre Technical Note No. 699)

Videotex Standard: Presentation Level Protocol, May 1981, Bell System

Government of Canada Department of Communications:

- Radio Standards Specification, RSS 151: "Low Power TV Broadcasting Transmitters Operating in the 54–88 MHz, 174–216 MHz, and 470–890 MHz Bands"
- Radio Standards Specification, RSS 154: "Television Broadcasting Transmitters Operating in the 54–88 MHz, 174–216 MHz, and 470–806 MHz Frequency Bands"
- Broadcast Specification, BS 13: "Ancillary Signals in the Vertical Blanking Interval for Television Broadcasting."

---

[3]Society of Motion Picture and Television Engineers Recommended Practice RP 27.3.

Appendix I.B

HAMMING CODE TABLE

ENCODING

INFORMATION BITS: b8, b6, b4, b2

HEXADECIMAL NUMBER	DECIMAL NUMBER	b8	b7	b6	b5	b4	b3	b2	b1
0	0	0	0	0	1	0	1	0	1
1	1	0	0	0	0	0	0	1	0
2	2	0	1	0	0	1	0	0	1
3	3	0	1	0	1	1	1	1	0
4	4	0	1	1	0	0	1	0	0
5	5	0	1	1	1	0	0	1	1
6	6	0	0	1	1	1	0	0	0
7	7	0	0	1	0	1	1	1	1
8	8	1	1	0	1	0	0	0	0
9	9	1	1	0	0	0	1	1	1
A	10	1	0	0	0	1	1	0	0
B	11	1	0	0	1	1	0	1	1
C	12	1	0	1	0	0	0	0	1
D	13	1	0	1	1	0	1	1	0
E	14	1	1	1	1	1	1	0	1
F	15	1	1	1	0	1	0	1	0

PROTECTION BITS: b7, b5, b3, b1

where

$b7 = b8 \oplus b6 \oplus b4$

$b5 = b6 \oplus b4 \oplus \overline{b2}$

$b3 = b4 \oplus \overline{b2} \oplus b8$

$b1 = \overline{b2} \oplus b8 \oplus b6$

DECODING

X1	X2	X3	X4	INTERPRETATION	INFORMATION
1	1	1	1	NO ERROR	ACCEPTED
0	0	1	0	ERROR IN b8	CORRECTED
1	1	1	0	ERROR IN b7	ACCEPTED
0	1	0	0	ERROR IN b6	CORRECTED
1	1	0	0	ERROR IN b5	ACCEPTED
1	0	0	0	ERROR IN b4	CORRECTED
1	0	1	0	ERROR IN b3	ACCEPTED
0	0	0	0	ERROR IN b2	CORRECTED
0	1	1	0	ERROR IN b1	ACCEPTED
$X1 \cdot X2 \cdot X3 = 0$			1	MULTIPLE ERRORS	REJECTED

where

$X1 = b8 \oplus b6 \oplus b2 \oplus b1$

$X2 = b8 \oplus b4 \oplus b3 \oplus b2$

$X3 = b6 \oplus b5 \oplus b4 \oplus b2$

$X4 = b8 \oplus b7 \oplus b6 \oplus b5 \oplus b4 \oplus b3 \oplus b2 \oplus b1$

Appendix I.C

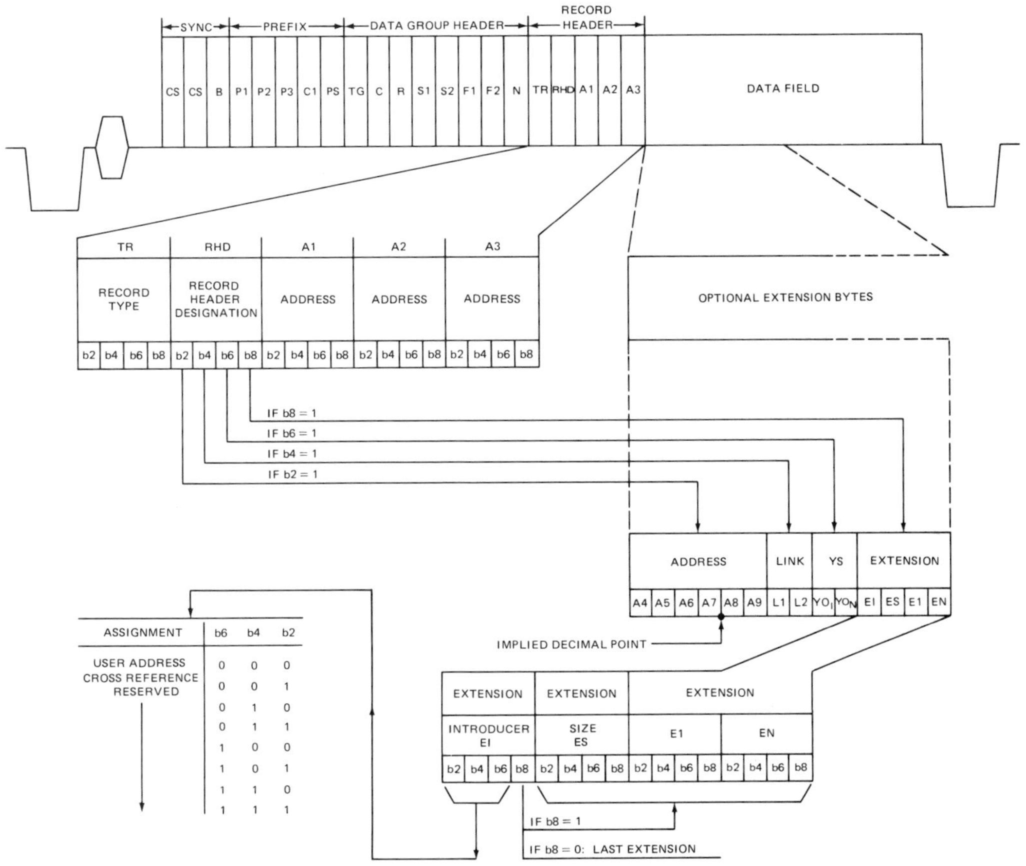
SYNC
PREFIX
DATA GROUP HEADER
RECORD HEADER
CS CS B P1 P2 P3 C1 PS TG C R S1 S2 F1 F2 N TR RHD A1 A2 A3
DATA FIELD
TR RHD A1 A2 A3
RECORD TYPE
RECORD HEADER DESIGNATION
ADDRESS
ADDRESS
ADDRESS
b2 b4 b6 b8 b2 b4 b6 b8 b2 b4 b6 b8 b2 b4 b6 b8 b2 b4 b6 b8
OPTIONAL EXTENSION BYTES
IF b8 = 1
IF b6 = 1
IF b4 = 1
IF b2 = 1
ADDRESS
LINK
YS
EXTENSION
A4 A5 A6 A7 A8 A9 L1 L2 YO1 YON EI ES E1 EN
IMPLIED DECIMAL POINT
ASSIGNMENT b6 b4 b2
USER ADDRESS 0 0 0
CROSS REFERENCE 0 0 1
RESERVED 0 1 0
0 1 1
1 0 0
1 0 1
1 1 0
1 1 1
EXTENSION INTRODUCER EI
EXTENSION SIZE ES
EXTENSION E1
EN
b2 b4 b6 b8 b2 b4 b6 b8 b2 b4 b6 b8 b2 b4 b6 b8
IF b8 = 1
IF b8 = 0: LAST EXTENSION

Appendix I.D

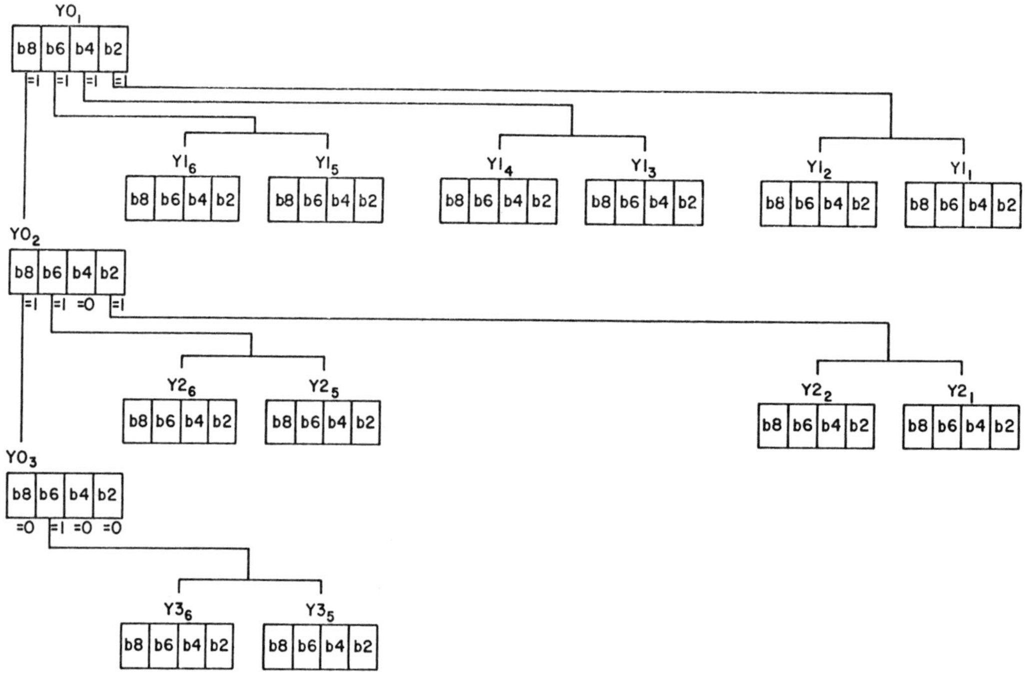

Order of transmission: YO_1 YI_1 YI_2 YI_3 YI_4 YI_5 YI_6 YO_2 $Y2_1$ $Y2_2$ $Y2_5$ $Y2_6$ YO_3 $Y3_5$ $Y3_6$

Appendix I.E

Interpretation of the Complementary Classification Bytes ($Y1_1 - Y1_6$, . . ., $YN_1 - YN_6$) and Header Extension Field Byte (EI).

This Appendix describes the recommended interpretation of the broadcast videotex standards. The specific assignments are chosen as working designations which will be adhered to indefinitely unless operating experience dictates otherwise.

Interpretation of Classification Bytes Associated with Byte YO_1

Byte $Y1_1$

b8,b6 Error Protection Levels: these bits are used to indicate the type of presentation level or error protection.

b8	b6	Four levels of message coding
0	0	Level 0 or no error protection being used
0	1	Level 1 or error protection being used
1	0	Reserved for future extension
1	1	Reserved for future extension

b4 = 1 Index message: This bit indicates that the message is an index message. Different index messages are possible which are numbered by byte $Y1_2$.

b2 = 1 Subindex: This bit indicates that the message is a submessage. Different subindexes are possible which are numbered by byte $Y1_2$.

Byte $Y1_2$

This byte is used to number index messages or subindex messages depending on the status of the bits in $Y1_1$.

Byte $Y1_3$

b8 = 1 Boxed message: This bit indicates that the message is related to the television program carrying the data signal. This bit indicates that the message is program related, and is to be displayed over the television program video with boxing. This may reset some options in the presentation layer to alternate values.

b6 = 1 Delayed message: The interpretation of the message and its presentation are delayed until the user manually requires its interpretation or until a specific "reveal" message is sent. This bit indicates that the message should not be revealed until a "reveal" message is sent, or until the user manually reveals the message.

b4,b2 Partial message: This indicates that the message contained in the Record cannot be interpreted alone. The terminal must first select a starting message which the partial message complements.

b4	b2	
0	0	Not partial message
0	1	First partial message
1	0	Last partial message
1	1	Intermediate partial message

Byte $Y1_4$

b8 = 1 Document Chain: This bit identifies a page which is part of a multipage Document (not last).

b6 = 1 Cyclic Marker: This bit identifies the first occurrence of any channel number in a cyclic information retrieval data base. This may be used by a decoder to abort a search for a requested page which is not present in the cycle.

b4 = 1 Auto Read: This bit indicates that a "Target Page" whose number is contained in the Extension Field which has its Header Extension code equal to "1," is to be captured immediately following the current one. This, however, requires user input (depression of 'proceed' key or equivalent) for display.

b2 = 1 Complementary Information: This bit indicates that complementary information is needed to properly interpret this message. The complementary information is to be found immediately following on the same Data Channel.

Byte $Y1_5$

b8 = 1 Program related message: This message is related to the television program carrying the data signal. When this bit is raised and the decoder is in the television mode, the message should be displayed over the television program.

b6 = 1 Alarm message: This bit indicates that the associated message has a priority function which can be interpreted by the decoder to override all other display functions.

b4 = 1 Update message: This is a flag indicating that the message contained in the Record replaces a previous message with the same address.

b2 = 1 New: This bit is used as an indicator to identify material not previously included in the information retrieval index. This permits decoders to be programmed to capture all new pages, or alternatively only those within a specific channel.

Byte $Y1_6$

Version: This byte, with four usage bits, is used to specify a version number of an information retrieval page.

Interpretation of Classification Bytes Associated with Byte $Y0_2$

Bytes $Y2_1$

Terminal: This byte, with four usage bits, is used to specify terminal functions.

Bytes $Y2_2 - Y2_6, \ldots, YN_1 - YN_6$

These bytes are not defined in this specification and are reserved for future extension.

Interpretation of Header Extension Field Byte (EI)

The first byte of the Header Extension Field is an extension field introducer byte (EI), with bit designations as follows:

b8 = 0 indicates last Extension Field

= 1 indicates further Header Extension Field(s) to follow ($E_1, \ldots, E_N$)

b6,b4,b2 indicate Header Extension Code assignments as follows:

= 0 reserved

= 1 cross reference;
identifies a Record to be
captured immediately following the current one.

= 2–7 reserved for future extension.

The second byte (ES) of the Header Extension Field indicates the number of bytes of Extension Field information in the current Extension Field.

AII

Line 21 Data Transmission Format

Extracted from *Telecaption Training Manual,* 1980. Used with the kind permission of Sears, Roebuck and Company.

Data Transmission Format

Captions associated with a television program will be transmitted as an encoded composite data signal during line 21 of field one of the standard NTSC video signal as shown in Figure 9. The signal consists of a clock run-in signal, a start bit, and 16 bits of data corresponding to two separate bytes of 8 bits each including parity. Therefore, transmission of actual data amounts to 16 bits every 1/30th of a second or 480 bits per second. This data stream contains encoded information that provides the instructions for display formating and the characters to be displayed.

The clock run-in consists of a seven-cycle sinusoidal burst that is frequency and phase locked to the caption data clock rate; the clock run-in signal whose frequency (64 fH) is twice that of the data clock provides synchronization for the decoder clock. The clock run-in signal is followed by two data bits at a zero logical level then a logical one-start bit. The last two cycles of the clock run-in, the two logical zero bits, and the logical one start bit constitute an eight-bit frame code signifying the start of data.

The transmitted data are coded in a nonreturn-to-zero (NRZ) format. All control and alphanumeric characters utilize the seven-bit code of the USA Standard Code for Information Interchange (USASCII). An eighth bit is added to each character to provide odd parity for error detection.

The sequence of identification, control and character transmission is shown in Figures 10–12. Each caption transmission is preceded by a preamble control code that consists

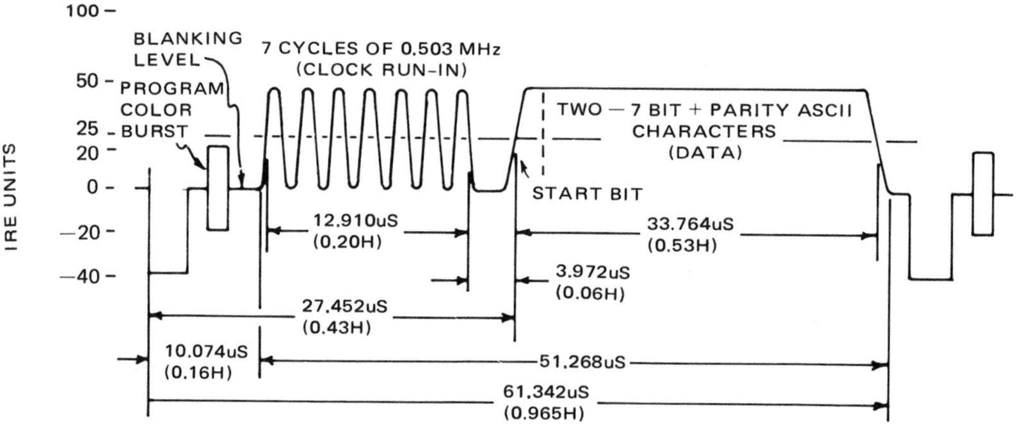

Figure 9.
Line 21 field 1 data signal format.

of a nonprinting character and a printing character to form a row address and display color code. Both characters of the preamble control code and all control codes are always transmitted within line 21 and twice in succession to insure correct reception of control information. Transmitted caption may be interrupted by midcaption control codes between two complete words in order to change display condition such as color or italics. At the completion of a caption transmission, an end of caption control code is sent.

Figure 10.
Caption row preamble format.

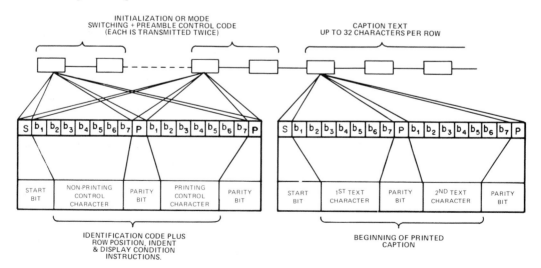

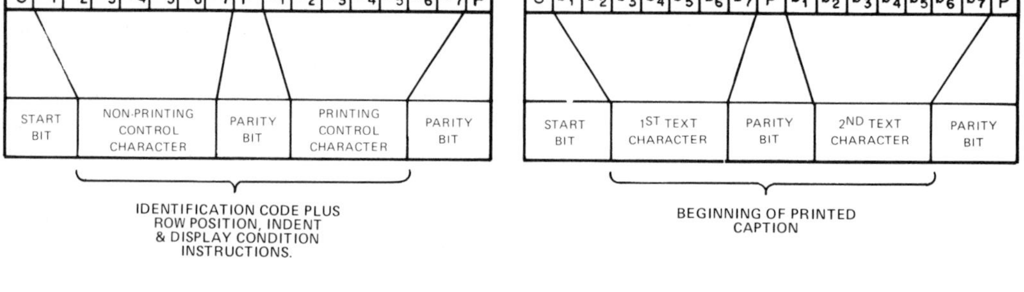

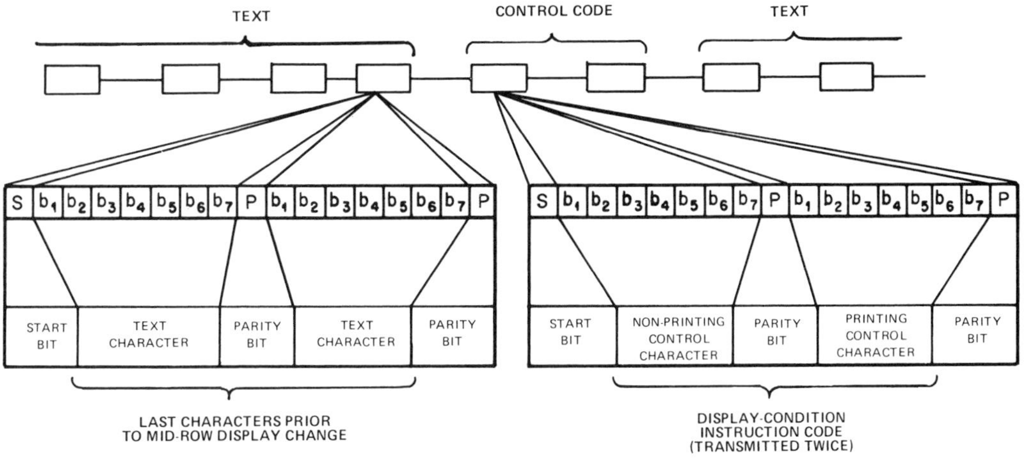

Figure 11.
Midcaption display—condition change format.

Figure 12.
End of caption and caption transition format.

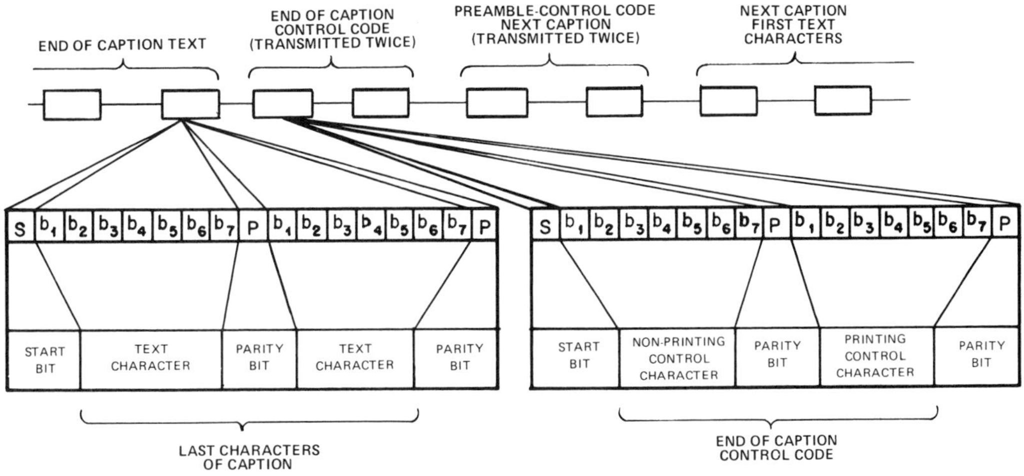

The first character of the control code is a nonprinting USASCII character (0000000 through 0011111) followed by a printing character (0100000 through 1111110). All characters that are received after a set of valid control codes are interpreted and loaded into memory as printing characters. Reception of any invalid control code will cause the system to ignore all subsequent character transmissions until receiving a valid control preamble. Character codes with bad parity result in an all ones code being written into memory; this causes display of a box (the delete symbol) in place of the desired character, which, of course, was in error.

NON-PRINTING CHARACTER b_7 b_6 b_5 b_4 b_3 b_2 b_1		PRINTING CHARACTER b_7 b_6 b_5 b_4 b_3 b_2 b_1	ROW NUMBER
0 0 1 0 0 0 1	(DC1)	1 0 A B C D E	1
0 0 1 0 0 0 1		1 1 A B C D E	2
0 0 1 0 0 1 0	(DC2)	1 0 A B C D E	3
0 0 1 0 0 1 0		1 1 A B C D E	4
0 0 1 0 0 1 1	(DC3)	1 0 A B C D E	12
0 0 1 0 0 1 1		1 1 A B C D E	13
0 0 1 0 1 0 0	(DC4)	1 0 A B C D E	14
0 0 1 0 1 0 0		1 1 A B C D E	15

A B C D	INDENT	CONDITION
0 0 0 0	0	MONOCHROME
0 0 0 1	0	GREEN
0 0 1 0	0	BLUE
0 0 1 1	0	CYAN
0 1 0 0	0	RED
0 1 0 1	0	YELLOW
0 1 1 0	0	MAGENTA
0 1 1 1	0	ITALICS
1 0 0 0	0	MONOCHROME
1 0 0 1	4	MONOCHROME
1 0 1 0	8	MONOCHROME
1 0 1 1	12	MONOCHROME
1 1 0 0	16	MONOCHROME
1 1 0 1	20	MONOCHROME
1 1 1 0	24	MONOCHROME
1 1 1 1	28	MONOCHROME

E = 1 UNDERLINE E = 0 NO UNDERLINE

Figure 15. (A) Catalog of control codes preamble control codes.

A valid control code always begins with one of four nonprinting ASCII control characters—DC1, DC2, DC3, or DC4—followed by a printing character code that when combined define all the required addressing and display functions. The complete catalog of these control codes is shown in Figure 15A, 15B.

Control Codes

The display controller operates in three basic data reception modes: normal caption, caption rolllup, and videotext. In each mode, each row of characters is always preceded by a preamble control code that specifies row color, underline, italics, indent, and row number. The very first character position of each written row is always displayed as a blank and the appropriate modified control code is written into its corresponding memory location. Midcaption control codes may be inserted within each row to change colors, start or stop underline, or switch to or from italics.

Normal captioning data are transmitted such that information is written into one caption memory while the other caption memory is displayed. On completion of a caption, the displayed memory may be erased and then the two memories are switched by the end of normal caption command. Thus the completed caption is "popped into" the TV screen and the other memory is ready to receive a new caption.

Caption rollup differs from the normal caption mode in that characters are written into the same memory that is being displayed. Two, three, or four rows of captions may be rolled up from row 15 at the bottom of the screen. The carriage return command initiates the 13-line rollup and triggers the single-line erase logic that erases the top row

NON-PRINTING CHARACTER b7 b6 b5 b4 b3 b2 b1	PRINTING CHARACTER b7 b6 b5 b4 b3 b2 b1	CONDITION
0 0 1 0 0 0 1	0 1 0 0 0 0 E	MONOCHROME
0 0 1 0 0 0 1	0 1 0 0 0 1 E	GREEN
0 0 1 0 0 0 1	0 1 0 0 1 0 E	BLUE
0 0 1 0 0 0 1	0 1 0 0 1 1 E	CYAN
0 0 1 0 0 0 1	0 1 0 1 0 0 E	RED
0 0 1 0 0 0 1	0 1 0 1 0 1 E	YELLOW
0 0 1 0 0 0 1	0 1 0 1 1 0 E	MAGENTA
0 0 1 0 0 0 1	0 1 0 1 1 1 E	ITALICS

E = 1 UNDERLINE E = 0 NO UNDERLINE

SPECIAL CHARACTERS

0 0 1 0 0 0 1	0 1 1 0 0 0 0	¼
0 0 1 0 0 0 1	0 1 1 0 0 0 1	/
0 0 1 0 0 0 1	0 1 1 0 0 1 0	½
0 0 1 0 0 0 1	0 1 1 0 0 1 1	¿
0 0 1 0 0 0 1	0 1 1 0 1 0 0	¾
0 0 1 0 0 0 1	0 1 1 0 1 0 1	¢
0 0 1 0 0 0 1	0 1 1 0 1 1 0	£
0 0 1 0 0 0 1	0 1 1 0 1 1 1	♪

MISCELLANEOUS CONTROL CODES

0 0 1 0 1 0 0	0 1 0 0 0 0 0	RESUME CAPTION LOADING WITHOUT RE-ADDRESSING
0 0 1 0 1 0 0	0 1 0 0 0 1 0	ALARM OFF
0 0 1 0 1 0 0	0 1 0 0 0 1 1	ALARM ON
0 0 1 0 1 0 0	0 1 0 0 1 0 1	ROLL-UP MODE - 2 ROWS
0 0 1 0 1 0 0	0 1 0 0 1 1 0	ROLL-UP MODE - 3 ROWS
0 0 1 0 1 0 0	0 1 0 0 1 1 1	ROLL-UP MODE - 4 ROWS
0 0 1 0 1 0 0	0 1 0 1 1 0 0	ERASE DISPLAYED MEMORY
0 0 1 0 1 0 0	0 1 0 1 1 1 1	END OF NORMAL CAPTION
0 0 1 0 1 0 0	0 1 0 1 0 1 0	DISPLAY FOLLOWING TEXT IMMEDIATELY STARTING IN ROW 1 POSITION 1
0 0 1 0 1 0 0	0 1 0 1 0 1 1	RESUME VIDEO TEXT LOADING WITHOUT RE-ADDRESSING
0 0 1 0 1 0 0	0 1 0 1 1 0 1	CARRIAGE RETURN

Figure 15. (B) Midcaption control codes.

12. New row characters are then written onto the previously erased row, which then appears on row 15 and rolls up with the other rows. The single-line erase occurs during line 238 immediately after receipt of the carriage return. Preamble control codes are still required at the beginning of each row but the row address is ignored because during rollup the logic is forced to write onto row 15.

Videotext is similar to caption rollup except the full 15 rows are utilized. Rollup is not initiated until 15 rows of information have been written. Receipt of the videotext restart command will cause erasure of videotext memory and new characters to be written onto row one starting with position one.

Character information may be multiplexed between the caption modes and videotext mode by use of the resume loading commands. Videotext may be transmitted and then a resume caption command sent. A previous caption may be completed or a new caption begun. A resume videotext command returns the logic to the videotext mode whereby additional character data may be sent starting with the last row and column position before the previous switch was made.

AIII

Interpretation Tables from the ANSI Videotex Teletext Presentation Level Protocol Syntax

There are 128 unique combinations of 1s and 0s when taken as a string of seven digits. In many computerized communication systems, characters are represented by 7-bit codes, with an eighth bit added as a parity bit to form an 8-bit byte. These 128 unique codes do not have to be interpreted the same way all the time, however. The interpretation can depend upon the use of a reference table to decide what a particular code will mean. In the ANSI standard based on the AT&T–suggested scheme for the interpretation of both 7-bit and 8-bit codes, there are a number of reference tables or interpretation tables that can be used. Even more reference tables can be defined in the future.

The first table, Table III.1, shows all the unique combinations of 1s and 0s for 7-bit "characters." The following interpretation tables are then used to decide what a 7-bit code means in a particular instance, and the indication of which table to use is itself determined by certain sequences of 7-bit codes transmitted to a receiver. Note that the first two columns always contain the same set of control codes, while the other six columns may contain different sets of characters depending upon which interpretation table is in use. In fact, a few of these other columns may even contain other control codes (e.g., when the "C1 Control Set" is the interpretation table for columns 4 and 5).

More complete details can be found in the draft standard of the American National Standards Institute titled *Videotex/Teletext Presentation Level Protocol Syntax (North American PLPS)*, and in the Bell System's *Presentation Level Protocol—Videotex Stan-*

Table III.1. Unique Codes for 7-Bit Characters

		Column							
		0	1	2	3	4	5	6	7
Row		000....	001....	010....	011....	100....	101....	110....	111....
0	...0000	0000000	0010000	0100000	0110000	1000000	1010000	1100000	1110000
1	...0001	0000001	0010001	0100001	0110001	1000001	1010001	1100001	1110001
2	...0010	0000010	0010010	0100010	0110010	1000010	1010010	1100010	1110010
3	...0011	0000011	0010011	0100011	0110011	1000011	1010011	1100011	1110011
4	...0100	0000100	0010100	0100100	0110100	1000100	1010100	1100100	1110100
5	...0101	0000101	0010101	0100101	0110101	1000101	1010101	1100101	1110101
6	...0110	0000110	0010110	0100110	0110110	1000110	1010110	1100110	1110110
7	...0111	0000111	0010111	0100111	0110111	1000111	1010111	1100111	1110111
8	...1000	0001000	0011000	0101000	0111000	1001000	1011000	1101000	1111000
9	...1001	0001001	0011001	0101001	0111001	1001001	1011001	1101001	1111001
10	...1010	0001010	0011010	0101010	0111010	1001010	1011010	1101010	1111010
11	...1011	0001011	0011011	0101011	0111011	1001011	1011011	1101011	1111011
12	...1100	0001100	0011100	0101100	0111100	1001100	1011100	1101100	1111100
13	...1101	0001101	0011101	0101101	0111101	1001101	1011101	1101101	1111101
14	...1110	0001110	0011110	0101110	0111110	1001110	1011110	1101110	1111110
15	...1111	0001111	0011111	0101111	0111111	1001111	1011111	1101111	1111111

dard, available through AT&T. Similar details are also provided in the CBS and the Telidon Videotex Systems filings with the Federal Communications Commission in the teletext proceedings.

With the exception of Table III.1, the tables have been reprinted from the ANSI draft standard dated June 18, 1982, and are used with permission.

Appendix III.A

Primary Character Set

					10	11	12	13	14	15
				b_7	0	0	1	1	1	1
				b_6	1	1	0	0	1	1
				b_5	0	1	0	1	0	1
b_4	b_3	b_2	b_1		2	3	4	5	6	7
0	0	0	0	0		0	@	P	`	p
0	0	0	1	1	!	1	A	Q	a	q
0	0	1	0	2	"	2	B	R	b	r
0	0	1	1	3	#	3	C	S	c	s
0	1	0	0	4	$	4	D	T	d	t
0	1	0	1	5	%	5	E	U	e	u
0	1	1	0	6	&	6	F	V	f	v
0	1	1	1	7	'	7	G	W	g	w
1	0	0	0	8	(	8	H	X	h	x
1	0	0	1	9	)	9	I	Y	i	y
1	0	1	0	10	*	:	J	Z	j	z
1	0	1	1	11	+	;	K	[	k	{
1	1	0	0	12	,	<	L	\	l	|
1	1	0	1	13	–	=	M	]	m	}
1	1	1	0	14	.	>	N	^	n	~
1	1	1	1	15	/	?	O	_	o	

Appendix III.B

Supplementary Character Set

b_4	b_3	b_2	b_1	b_7 b_6 b_5	10 (0 1 0) 2	11 (0 1 1) 3	12 (1 0 0) 4	13 (1 0 1) 5	14 (1 1 0) 6	15 (1 1 1) 7
0	0	0	0	0		°	→	—	Ω	ĸ
0	0	0	1	1	¡	±	`	¹	Æ	æ
0	0	1	0	2	¢	²	´	®	Đ	đ
0	0	1	1	3	£	³	^	©	ª	ð
0	1	0	0	4	$	×	~	T.M.	Ħ	ħ
0	1	0	1	5	¥	µ	¯	♪	⊞	ı
0	1	1	0	6	#	¶	˘	⊟	IJ	ij
0	1	1	1	7	§	•	˙	◫	Ŀ	ŀ
1	0	0	0	8	¤	÷	¨	⧄	Ł	ł
1	0	0	1	9	‘	’	/	⧅	Ø	ø
1	0	1	0	10	“	”	˚	◪	Œ	œ
1	0	1	1	11	«	»	¸	◩	º	ß
1	1	0	0	12	←	¼	▯	⅛	Þ	þ
1	1	0	1	13	↑	½	˝	⅜	Ŧ	ŧ
1	1	1	0	14	→	¾	˛	⅝	Ŋ	ŋ
1	1	1	1	15	↓	¿	ˇ	⅞	ŉ	

Appendix III.C

PDI Code Sets

					10	11	12	13	14	15
				b7	0	0	1	1	1	1
				b6	1	1	0	0	1	1
				b5	0	1	0	1	0	1
					2	3	4	5	6	7
b4	b3	b2	b1							
0	0	0	0	0	RESET	RECT (OUT-LINED)				
0	0	0	1	1	DOMAIN	RECT (FILLED)				
0	0	1	0	2	TEXT	SET & RECT OUT-LINED)				
0	0	1	1	3	TEXTURE	SET & RECT (FILLED)				
0	1	0	0	4	POINT SET (ABS)	POLY (OUT-LINED)				
0	1	0	1	5	POINT SET (REL)	POLY (FILLED)				
0	1	1	0	6	POINT (ABS)	SET & POLY (OUT-LINED)				
0	1	1	1	7	POINT (REL)	SET & POLY (FILLED)	NUMERIC DATA			
1	0	0	0	8	LINE (ABS)	FIELD				
1	0	0	1	9	LINE (REL)	INCR POINT				
1	0	1	0	10	SET & LINE (ABS)	INCR LINE				
1	0	1	1	11	SET & LINE (REL)	INCR POLY (FILLED)				
1	1	0	0	12	ARC (OUT-LINED)	SET COLOR				
1	1	0	1	13	ARC (FILLED)	WAIT				
1	1	1	0	14	SET & ARC (OUT-LINED)	SELECT COLOR				
1	1	1	1	15	SET & ARC (FILLED)	BLINK				

Appendix III.D

Mosaic Set

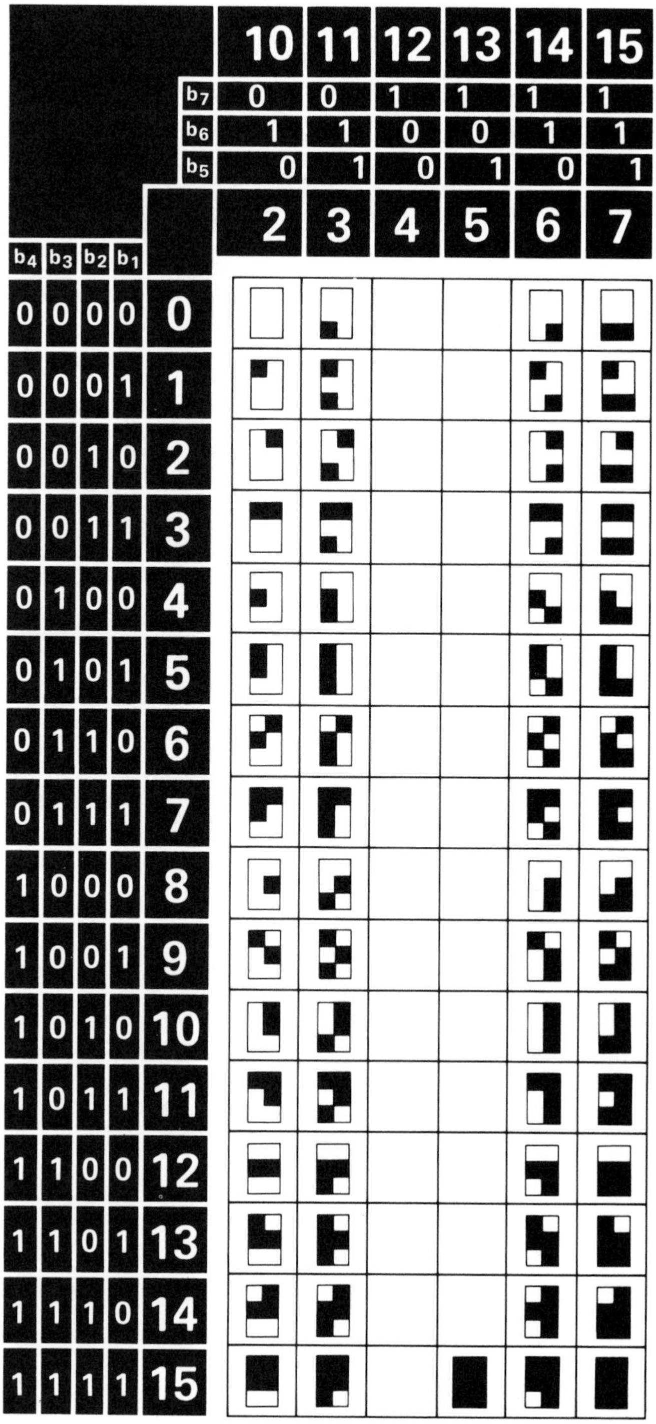

10 11 12 13 14 15
b7 0 0 1 1 1 1
b6 1 1 0 0 1 1
b5 0 1 0 1 0 1
2 3 4 5 6 7
b4 b3 b2 b1
0 0 0 0 0
0 0 0 1 1
0 0 1 0 2
0 0 1 1 3
0 1 0 0 4
0 1 0 1 5
0 1 1 0 6
0 1 1 1 7
1 0 0 0 8
1 0 0 1 9
1 0 1 0 10
1 0 1 1 11
1 1 0 0 12
1 1 0 1 13
1 1 1 0 14
1 1 1 1 15

Appendix III.E

C0 Control Set

b_7					0	0
b_6					0	0
b_5					0	1
b_4	b_3	b_2	b_1	COLUMN / ROW	0	1
0	0	0	0	0	NUL	DLE
0	0	0	1	1	SOH	DC1
0	0	1	0	2	STX	DC2
0	0	1	1	3	ETX	DC3
0	1	0	0	4	EOT	DC4
0	1	0	1	5	ENQ	NAK
0	1	1	0	6	ACK	SYN
0	1	1	1	7	BEL	ETB
1	0	0	0	8	APB (BS)	CAN
1	0	0	1	9	APF (HT)	SS2
1	0	1	0	10	APD (LF)	SUB
1	0	1	1	11	APU (VT)	ESC
1	1	0	0	12	CS (FF)	APS
1	1	0	1	13	APR (CR)	SS3
1	1	1	0	14	SO	APH
1	1	1	1	15	SI	NSR

Appendix III.F

C1 Control Set

Definitions of Columns A and B. 1) If a C1 control function is represented by a 2-character escape sequence (in a 7-bit code), the table specifies the bit combination of the final character by taking A = 4 and B = 5. 2) If a C1 control function is represented by a single 8-bit combination the table specifies this combination by taking A = 8 and B = 9.

b_4	b_3	b_2	b_1			
				b_8	1	1
				b_7	0	0
					8	9
				b_7	1	1
				b_6	0	0
				b_5	0	1
					4	5
				COLUMN / ROW	A	B
0	0	0	0	0	DEF MACRO	PROTECT
0	0	0	1	1	DEFT MACRO	EDC1
0	0	1	0	2	DEFT MACRO	EDC2
0	0	1	1	3	DEF DRCS	EDC 3
0	1	0	0	4	DEF TEXTURE	EDC 4
0	1	0	1	5	END	WORD WRAP OFF
0	1	1	0	6	REPEAT	WORD WRAP ON
0	1	1	1	7	REPEAT TO EOL	SCROLL ON
1	0	0	0	8	REVERSE VIDEO	SCROLL OFF
1	0	0	1	9	NORMAL VIDEO	UNDER LINE START
1	0	1	0	10	SMALL TEXT	UNDER LINE STOP
1	0	1	1	11	MED TEXT	FLASH CURSOR
1	1	0	0	12	NORMAL TEXT	STEADY CURSOR
1	1	0	1	13	DOUBLE HEIGHT	CURSOR OFF
1	1	1	0	14	BLINK START	BLINK STOP
1	1	1	1	15	DOUBLE SIZE	UNPROTECT

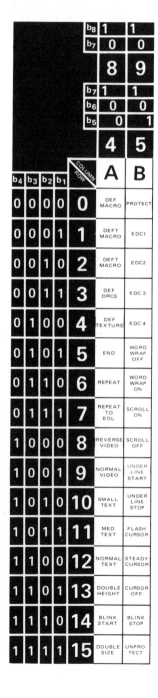

Selected Bibliography

Acceptance of Videotex Uncertain. *ASIS Bulletin,* December 1981, p. 14.

Akgun, Metin B. Comparison of Technology and Capital Costs for New Home Services. *IEEE Transactions on Cable Television* CATV–5 (3): 124–138 (July 1980).

All Together Now On Teletext. *Broadcasting,* May 18, 1981, pp. 28–29.

Alternate Media Center, New York University. Research on Broadcast Teletext: Working Paper Number One. Access Time and Reception Quality in the Field Trial in Washington, DC. September 1981.

______. Research on Broadcast Teletext: Working Paper Number Three. Early Use of Graphics in the Field Trial in Washington, DC. December 1981.

______. Research on Broadcast Teletext: Working Papers Number Four and Five, February 1982.

______. Teletext and Public Broadcasting. April 1980.

American National Standards Institute, Subcommittee X3L2.1. Videotex/Teletext Presentation Level Protocol Syntax (North American PLPS), draft standard, June 1982.

American Newspaper Publishers Association, Telecommunications Committee. Information Systems, March 1981, 40 pp.

Arlen, Gary, ed. *International Videotex Teletext News,* 1980– (monthly newsletter).

______. The Revenue Potential of Teletext. *View,* June 1981, pp. 90–94.

Baker, Jeri. Information Services: Reading Between the Lines. *Cablevision,* August 10, 1981, pp. 32–37.

______. Must Reading? *Cabelvision* (Plus), March 8, 1982, pp. 4–10.

______. Out of the Time Incubator. *Cablevision* (Plus), March 22, 1982, pp. 17–18.

Bell System. *Presentation Level Protocol. Videotex Standard.* Parsippany, N.J.: American Telephone and Telegraph Company, 1981.

Benton and Bowles, Inc. The New TV Technologies: The View from the Viewer. An American Consensus Report, March 1981, 71 pp.

Bown, H. G., O'Brien, C. C., Sawchuk, W.; and Storey, J. R. A Canadian Proposal for Videotex Systems: General Description. Communications Research Centre, November 1978, 130 pp.

Bown, Herbert G., and Sawchuk, William. Telidon—A Review. *IEEE Communications Magazine,* January 1981, pp. 22–28.

Burnham, David. The Twists in Two-Way Cable. *Channels,* June–July 1981, pp. 38–44.

CBS, Inc. *North American Broadcast Teletext Specification,* June 28, 1981.

———. Reply by CBS Inc. to Statements Concerning Petition for Rulemaking of the United Kingdom Teletext Industry Group. Before the Federal Communications Commission, July 21, 1981.

CBS Presses FCC for Teletext Standard. *Broadcasting,* January 18, 1982, pp. 80–81.

CCIR Working Group II. Additional Broadcasting Services Using a Television or Narrowband Channel. Report 802 (Mod I), October 8, 1981.

———. Data Broadcasting Systems—Signal and Service Quality, Field Trials and Theoretical Studies, draft report, AE/11, October 3, 1981.

CEPT Sub-Working Group CD/SE. European Interactive Videotex Service, Display Aspects and Transmission Coding, undated, 64 pp.

CSP International. Market Analysis: Teletext in the United States, October 1981, 51 pp.

Cable '81 Technical Papers. Proceedings of the 30th Annual Convention of the National Cable Television Association. Washington, D.C.: National Cable Television Association, 1981.

Canada, Department of Communications. *Broadcast Specification, Provisional. Television Broadcast Videotex,* June 19, 1981.

Champness, Brian G., and de Alberdi, Marco. Measuring Subjective Reactions to Teletext Page Design. Alternate Media Center, New York University, September 1981.

Chouinard, Martin. Television on Request Via Multiple Channel Data Display Systems. *Cable Communications,* May 1981, pp. 42–45.

Ciciora, Walter S. Teletext Systems: Considering the Prospective User. *SMPTE Journal* 89(11): 846–849 (November 1980).

———. Twenty-Four Rows of Videotex in 525 Scan Lines, unpublished paper, Zenith Radio Corporation, 1981.

———. Virtext and Viewdata: Adventures in Vertical Interval Signaling. *Cable '81,* Proceedings of the 30th Annual Convention of the National Cable Television Association, Washington, D.C.: National Cable Television Association, 1981, pp. 101–104.

Consumer Text Display Systems (Teletext and Viewdata). *IEEE Transactions on Consumer Electronics* (Special Issue) CE–25(3): entire issue (July 1979).

Dages, Charles L. Playcable: A Technological Alternative for Information Services. *IEEE Transactions on Consumer Electronics* CE–26(3): 482–486 (August 1980).

De Sonne, Marcia. Leasing Your SCA—A New Business Opportunity. *RadioActive* 8(6): 8–10 (June–July 1982).

Edwards, Kenneth. Teletext Broadcasting in U.S. Endorsed by FCC. *Editor and Publisher,* November 18, 1978, pp. 11–12.

Edwards, Morris. Videotex/Teletext Services Seek Haven in USA Homes. *Communications News,* August, 1980, pp. 38–41.

Electronic Industries Association, Broadcast Television Systems Committee, Teletext Subcommittee. *Interim Report,* vols. I and II. Washington, D.C.: Electronic Industries Association, 1981.

European Interactive Videotex Service, Display Aspects and Transmission Coding. British Telecom Research Laboratories, 1981.

Federal Communications Commission. FCC to Consider Authorization of Teletext Service by TV Stations, BC Docket No. 81–741, *FCC News,* October 22, 1981.

Fedida, Sam, and Malik, Rex. *Viewdata Revolution.* New York: Halsted Press, 1979.

Feldman, Robert. FCC Considers Stepping Aside in Teletext-Videotex Standards Fight. *Electronic Engineering Times,* August 17, 1981, pp. 1, 10–11.

Fuller, Keith. Report on AP/Newspaper/Compuserve Experiment, Associated Press, June 1981, 7 pp.

Goldman, Ronald J. Teletext for the Home and School at KCET. *E-ITV,* October 1981, pp. 39–41.

Harris, A. B. Guide to Videotex Presentation Level Standards, British Telecom, July 1981.

Here Comes Talkie-Text; Difficult Digitizing Problems; AT&T Intends Test with Knight-Ridder. *Communications Daily,* November 16, 1981, pp. 1–2.

Herring, William. Teletext User Survey. Ceefax and Oracle—User Reactions, Washington, D.C.: Videotex Industry Association, 1982.

Hill, I. William. Seek Editor Input on Future of Video Screen Newspapers. *Editor & Publisher,* September 4, 1976, p. 14.

The Home Information Revolution. *Business Week,* June 29, 1981, pp. 74–83.

Howard, Niles. The Looming Battle Over Videotext. *Dun's Review,* February 1981, pp. 58–63.

Inside Videotex. Proceedings of a seminar, March 13–14, 1980. Toronto: Infomart, 1980.

Ishigaki, Y., Okada, Y., Hashimoto, T., and Ishikawa, T. Television Design Aspects for Better Teletext Reception. *IEEE Transactions on Consumer Electronics CE–26(3) 622–628 (August 1980).*

Jackson, Charles L., Shooshan, Harry M., and Wilson, Jane L. *Newspapers and Videotex: How Free a Press?* St. Petersburg, Fla.: Modern Media Institute, 1981.

Jackson, Richard N. Home Communications I: Teletext and Viewdata. *IEEE Spectrum,* March 1980, pp. 26–32.

Japan, Ministry of Posts and Telecommunications; and Nippon Telegraph and Telephone Public Corporation. *Captain System. Character and Pattern Telephone Access Information Network System,* undated.

Lavers, Daphne. International Videotex Conference and Exhibition Report. *Cable Communications,* July 1981, pp. 52–57.

Lentz, John, Sillman, David, Thedick, Henry, and Wetmore, Evans. Television Captioning for the Deaf. Signal and Display Specifications. Report no. E–7709–C, Engineering and Technical Operations Department, Public Broadcasting Service, May 1980, revised.

Levesque, Pierre. Comments on PDI Usage in CBC Teletext Planning. Canadian Broadcasting Corporation, 1981.

Lilley, William, et al. *New Technologies Affecting Radio and Television Broadcasting*. Washington, D.C.: National Association of Broadcasters, 1981.

Link Resources Corporation. *Viewdata/Videotex Report,* 1980– (monthly newsletter).

Lipoff, Stuart J. Data Analysis Procedures and Presentation Formats for Broadcast Teletext Field Tests. *IEEE Transactions on Consumer Electronics* CE–26(3): 507–518 (August 1980).

Lopinto, John. The Application of DRCS Within the North American Broadcast Teletext Specification. Time, Inc., November 17, 1981.

______. Establishing Practices in International Teletext. *CED* (Communications Engineering Digest), December 1981, pp. 30–38.

A Lot of Interest in Teletext at 59th Gathering of the NAB. *Variety,* April 14, 1981, p. 8.

McGee, William L., and Garrick, Lucy E. Videotex—Teletext, Viewdata, Cable Text. In *Changes, Challenges and Opportunities in the New Electronic Media*. San Francisco: Broadcast Marketing Company, 1982.

Mahoney, Sheila, DeMartino, Nick, and Stengel, Robert. *Keeping Pace with the New Television*. New York: VNU Books, International, 1980.

Microband Corporation of America. Proposal of Microband Corporation of America for the Creation of Urban Over-the-Air "Wireless Cable" Networks Capable of Providing Premium Television and Other Broadband and Narrowband Communications Services. Before the Federal Communications Commission, undated (1982).

NCTA Opposes Adoption of British Teletext. *CED* (Communications Engineering Digest), June 1981, p. 13.

National Association of Broadcasters. Videotex—Teletext—Viewdata. A Selective Bibliography. January 1981, 4 pp.

Neustadt, Richard M. *The Birth of Electronic Publishing*. White Plains, N.Y.: Knowledge Industry Publications, 1982.

Neustadt, Richard M., Skall, Gregg P., and Hammer, Michael. The Regulation of Electronic Publishing. *Federal Communications Law Journal* 33(3): 331–417 (Summer 1981).

Newspapers Get a Glimpse of Their Electronic Future. *Broadcasting,* February 1, 1982, pp. 68–69.

Nicholls, William C., and Seidel, Robert P. Comments on ANSI Draft Standard for Videotex/Teletext Presentation Level Protocol Syntax from a Teletext Viewpoint. CBS Television Network Engineering and Development Department, CBS Inc., June 1982.

1982 Teletext Stampede Seen by CBS Executive. *Broadcast Management/Engineering,* January 1982, pp. 12–13.

Panero, Hugh. Dow Jones' Double Cable Slice: Information Services, System Ownership. *Cablevision,* March 9, 1980, pp. 45–49.

Philips Electronic and Associated Industries, Ltd. Teletext and the Consumer. August 1981, 20 pp.

Public Stations Opening Up to Closed Captioning. *Broadcasting,* June 29, 1981, pp. 64–65.

RCA Corporation. Comments of RCA Corporation. Before the Federal Communications Commission in the matter of Amendment of Part 73 to Authorize the Transmission of Teletext by TV Stations, February 5, 1982, 6 pp.

Reilly, Michael J. Generating Teletext Characters by Row-Grabbing Technique. *CED* (Communications Engineering Digest), December 1981, pp. 20–22.

Roizen, Joe. The Promise of Teletext Is Nearing the Marketplace. *Broadcast Communications,* December 1981, pp. 53–55.

———. Teletext Tests Will Make More Than News. *Broadcast Communications,* April 1981, pp. 38–41.

Rothbart, Gary. Screening the Future. *Cablevision* (Plus), March 8, 1982, pp. 15–17.

Sawchuk, W., Bown, H. G., O'Brien, C. D., and Thorgeirson, G. W. An Interactive Image Communications System Using Narrowband Lines. *Computers and Graphics 3: 129–134 (1978).*

Schober, Gary. The WETA Teletext Field Trial: Some Technical Concerns. Paper presented at the Spring Conference of the IEEE on Consumer Electronics, June 3–4, 1981.

Schuyten, Peter J. The Dispute Over Teletext. *The New York Times,* August 28, 1980, p. D2.

Sigel, Efrem (ed.). *Videotext.* White Plains, N.Y.: Knowledge Industry Publications, Inc., 1980.

Signetics Corporation. Integrated Circuits for Videotex and Teletext. Brochure, 1981, 6 pp.

Simons, David M. Cable Courts Videotex. *ASIS Bulletin,* December 1981, p. 32.

Somers, Eric. Appropriate Technology for Text Broadcasting. *Videotex, Viewdata & Teletext.* Transcript of the Online Conference on Videotex, Viewdata and Teletext. Northwood Hills, England: Online Conferences, Ltd., 1980, pp. 499–514.

———. The Growing Potential of FM/SCA. *Radioactive* 5(2): 8–9 (February 1979).

Southern Satellite Systems. Cabletext Specifications and Options. December 16, 1981, 3 pp.

Storey, J. R., Bown, H. G., O'Brien, C. D., and Sawchuk, W. An Overview of Broadcast Teletext Systems for NTSC Television Standards. Unpublished paper, Department of Communications, Ottawa, Canada, 21 pp.

Storey, J. R., Vincent, A., and FitzGerald, R. A Description of the Broadcast Telidon System. Paper presented at the IEEE Chicago Spring Conference on Consumer Electronics, June 18–19, 1980.

Sullivan, Bill. Specialized Uses Can Make Teletext Services Viable Now. *TVC,* April 1, 1981, pp. 40–42.

Telecaption Training Manual. Chicago: Sears, Roebuck & Company, 1980.

Teletext and Videotext. *Broadcasting,* June 28, 1982, pp. 37–49.

Teletext and Viewdata Services. Proceedings of the 1980 Spring Engineering Conference of the Society for Cable Television Engineers. Washington, D.C.: Society of Cable Television Engineers, 1980.

Teletext Standard Put Before FCC by CBS and TVS. *Broadcasting,* July 27, 1981, pp. 102–103.

Telidon Videotex Systems, Inc. Comments of Telidon Videotex Systems, Inc. Before the Federal Communications Commission in the matter of Amendment of Part 73 to Authorize the Transmission of Teletext by TV Stations, June 8, 1981.

Treurniet, W. C. Display of Text on Television. CRC Technical Note no. 705–E. Ottawa, Canada: Communications Research Centre, 1981.

Tydeman, J., Lipinski, H., Adler, R., Nyhan, M., and Zwimpfer, L. *Teletext and Videotex in the United States*. Menlo Park, Calif. Institute for the Future, 1982. Also published under the same title by McGraw-Hill Publications Company, 1982.

United Kingdom Teletext Industry Group. Petition for Rule-making of United Kingdom Teletext Industry Group. Before the Federal Communications Commission in the matter of Amendment of Part 73 to Authorize the Transmission of Teletext by TV Stations, March 26, 1981.

______. Reply Comments of United Kingdom Teletext Industry Group. Before the Federal Communications Commission, July 21, 1981.

Vermilyea, David, and Wylie, Donald. Teletext in the Year 2000: A Delphi Forecast. Proceedings of the National Telecommunications Conference of the IEEE. New York: The Institute of Electrical and Electronics Engineers, Inc., 1980, pp. 23.4.1–23.4.4.

Videotex 81. Proceedings of a conference. Northwood Hills, England: Online Conferences, Ltd., 1981.

Videotex—Key to the Information Revolution. Proceedings of Videotex 82. Northwood Hills, England: Online Conferences, Ltd., 1982.

Viewdata 80. Proceedings of the First World Conference on Viewdata, Videotex and Teletext. Northwood Hills, England: Online Conferences, Ltd., 1980.

Viewdata 81. Proceedings of a conference. Northwood Hills, England: Online Conferences, Ltd., 1981.

Viewtext 81.

Proceedings of a conference, April 23–24, 1981. Brookline, Mass.: Information Gatekeepers, Inc., 1981.

Vivian, R. H. Level 4 Enhanced UK Teletext Transmits Graphics through Efficient Alpha-geometric Coding. Independent Broadcasting Authority, 1982.

Wells, Daniel R. PBS Captioning for the Deaf. Engineering Report no. E–7907, Public Broadcasting Service, 1979.

Who's Doing What with Teletext? E–ITV, October 1981, pp. 47–49.

Window on the World. The Home Information Revolution. *Business Week,* June 29, 1981, pp. 74–83.

Woolfe, R. *Videotex. The New Television-Telephone Information Services*. London: Heyden & Son, Ltd., 1980.

Wright, William F., and Hawkins, Donald T. Information Technology, a Bibliography. *Special Libraries* 72(2): 163–173 (April 1981).

Zenith Radio Corporation. Comments of Zenith Radio Corporation. Before the Federal Communications Commission, August 27, 1981.

______. Comments of Zenith Radio Corporation. Before the Federal Communications Commission in the Matter of Amendment of Part 73 to Authorize the Transmission of Teletext by TV Stations, February 9, 1982.

Index

11-6-84

[redacted]